# INTRODUCTION TO INDOOR FISH FARMING

# INTRODUCTION TO INDOOR FISH FARMING

*Edited by*

**Dr. Dhanaji Patil**

Delve PUBLISHING

www.delvepublishing.com

**Introduction to Indoor Fish Farming**

*Dr. Dhanaji Patil*

**Delve Publishing**
224 Shoreacres Road
Burlington, ON L7L 2H2
Canada
www.delvepublishing.com
Email: orders@arclereducation.com



ISBN: 978-1-77469-838-9 (Hardcover)

# ABOUT THE EDITOR

**Dr. Dhanaji W. Patil** is serving as Head and Assistant Professor in Department of Fishery Science, Toshniwal Arts, Commerce and Science College, Sengaon, Dist: Hingoli, Maharashtra. He completed his graduation in Bachelor of Fishery Science (B. F. Sc.) during 2003 and Master of Fishery Science (M. F. Sc.) during 2005 from Dr. Balasaheb Sawant Konkan Krishi Vidyapeeth, Dapoli (MS). He has qualified NET (Fishery Science) conducted by National Agriculture Recruitment Board (ICAR), New Delhi in 2012. He completed Ph. D. in Fishery Science from Swami Ramanand Teerth Marathwada University, Nanded, Maharashtra in 2020. He joined as Assistant Professor on October 2012 and his area of specialization is Aquaculture. He has field experience especially in field of Aquaculture prior to joining. He has completed two research projects funded by Rajiv Gandhi Science and Technology Commission, Mumbai. He has published 19 Research articles along with 02 book chapters and 05 extension publications. He organized one day National Conference on "Advances in Fisheries, Biological and Allied Research (AFBAR-2019)" during March 2019 as a Convener. He organized and conducted 04 training programs on aquaculture and fish seed production technologies for farmers and fishermen, 02 faculty development programs on Educational Video Creation- E content Development for teaching community all over India. He contributed as resource person in 14 Faculty Development Programmes conducted by different institutions of the country. He has also participated in several seminars and conferences and present his research work. He has developed E-content through YouTube Channel (Fisheries Only). He is also member of Asian Fisheries Society.

# TABLE OF CONTENTS

# LIST OF FIGURES

# LIST OF ABBREVIATIONS

| | |
|---|---|
| AHP | analytical hierarchy process |
| ASC | Aquaculture Stewardship Council |
| CFP | common fisheries policy |
| CNS | central nervous system |
| COC | cumulative oxygen consumption |
| DHA | docosahexaenoic acid |
| DIN | dissolved inorganic nitrogen |
| DL-PCB | dioxin-like PCBs |
| EIA | environmental impact assessment |
| EPA | eicosapentaenoic acid |
| FPH | fish protein hydrolysate |
| HPLC | high pressure liquid chromatography |
| HVAC | heating, ventilation and air-conditioning |
| IMTA | integrated multi-trophic aquaculture |
| IPM | integrated pest management |
| K | potassium |
| MDG | millennium development goal |
| MIC | minimum inhibitory concentration |
| MPA | marine protected area |
| MSY | maximum sustainable yield |
| N | nitrogen |
| NFT | nutrient film technique |
| NGS | next generation of sequencing |
| $NH_4$ | ammonium |
| $NO_3$ | nitrate |

| | |
|---|---|
| OTC | oxytetracycline |
| PAM | pralidoxime |
| PB1 | polar bodies I |
| PCBs | polychlorinated biphenyls |
| PCDD | polychlorinated dibenzo-para dioxins |
| PCDF | polychlorinated dibenzofurans |
| POPs | persistent organic pollutants |
| ppt | parts per thousand |
| PRAs | participatory rural appraisal |
| PUFA | polyunsaturated fatty acids |
| PVC | polyvinyl chloride |
| RAS | recirculating aquaculture system |
| RFLP | restriction fragment length polymorphism |
| RH | radiation hybrid |
| SAGE | serial analysis of gene expression |
| SDD | Secchi disc depth |
| SLA | sustainable livelihoods approach |
| SNPs | single nucleotide polymorphisms |
| SSF | small-scale fisheries |
| SSRs | simple sequence repeats |
| TACs | total permitted catches |
| TBT | tributyltin |
| TDI | tolerated daily intakes |
| TEF | toxic equivalent factors |
| TEQ | toxic equivalents |
| TIN | total inorganic nitrogen |
| TSS | total suspended solids |
| UV | ultraviolet |
| WHO | World Health Organization |

# PREFACE

The farming of aquatic animals in house premises is known as indoor fish farming or indoor aquaculture. Its major goal is to produce aquatic food species for human consumption, but it also has other goals, such as ornamental and aquarium fish cultivation. Aquaculture can be done on a variety of scales. It ranges from small rainfed ponds to big commercial farms to boost food production for rural as well as urban families to supply export items for worldwide markets. Aquaculture is an economic activity, regardless of scale and the value of aquacultural crops must surpass the cost of production. Because our understanding of the processes and interactions that impact success in aquaculture is less developed than in traditional agriculture, aquaculture has tended to be riskier and, in some cases, unsustainable. However, aquaculture technology is continuously advancing, and more reliable production techniques are becoming available.

Indoor fish farming is one of the oldest human activities, and fishing equipment has been invented. Inland and indoor fish has remained an important source of animal protein to this day, especially in locations far from the sea. The current high demand for fish, along with growing knowledge of the environment's significance in human well-being, has resulted in a rapid shift in attitudes about inland water resources. This shift is part of a greater concern for ecosystem long-term stability, as well as the capacity of remaining freshwater resources to meet human demands. Humanity is shifting away from old disruptive development patterns and focusing on agriculture, forestry as well as fisheries as priorities in order to ensure the long-term viability of the world's resources and ensure long-term food security. Conservation and sustainability policies are being adopted in the face of growing constraints on natural resources in general. These increases in resource consumption and environmental impact are most visible in inland waters, which are especially susceptible since they operate as collectors of all activities taking place in their basins and are among the world's most endangered ecosystems. Indoor fisheries and aquaculture are heavily influenced by the social and geographical setting in which they are pursued, and they adapt to the current world's demographic, economic, and political evolution.

This book comprehensively focuses on introductory concepts of indoor fish farming and its impact on human beings. This book serves as a manual for students pursuing graduation, post graduate and research studies.

# Introduction

Nutrient concentrations rise as the intensity of aquaculture is raised by the use of manures, fertilizers, and feeds, plankton blooms occur, and dead organic debris sinks to pond bottoms. This is common in case of indoor fish farms. These farms have phytoplankton that generates dissolved oxygen through photosynthesis; however, as the biomass of pond biota increases in response to nutrient inputs, respiration increases as well. Anaerobic conditions and the release of microbial metabolites such as nitrite and hydrogen sulfide into the water column can result from the accumulation of organic debris on pond bottoms. The toxins released from these activities are harmful to fishes and leading to death also. Ammonia is the primary nitrogenous waste product of aquaculture species, and when production intensifies, ammonia concentrations may reach lethal levels. Low dissolved oxygen concentrations are common in indoor fish farming, on the other hand, are frequently the initial limiting factor in indoor fish production.

Production can be raised until ammonia toxicity begins to limit production if mechanical aeration is supplied to increase dissolved oxygen content. Water exchange is the most reliable method of lowering ammonia levels. Production levels could be boosted by exchanging water to discharge ammonia from indoor ponds. Because microbial decomposition of stored organic waste at the base of the indoor fish pond in deterioration of bottom soil condition and high concentrations of toxic metabolites, production cannot be enhanced indefinitely through aeration and water exchange. The water quality in fish ponds is a crucial determinant of fish output. Indoor fish farm management strategies such as fish stocking density, fertilization strategy, and supplemental feeding have a significant impact on water quality. Chemical or physical methods such as aeration, fertilizing, liming, or water exchange can be used to adjust water quality in an attempt to overcome pond production restrictions.

The urban areas are more inclined to install indoor fish farming equipment and this is spreading to all countries. This can be attributed due to the inherent benefits offered in this system with limited costs. It can also create employment opportunities for several people without leaving their native places. Though there are several factors that are to be considered while installing and operating these farms, yet this is profitable. This book provides an elementary understanding of the indoor fish farming practices and challenges that are to be faced by the farmers.

CHAPTER 1

# Introduction to Indoor Fish Farming

CONTENTS

## 1.1. INTRODUCTION

Engineers have been struggling to keep up with the current codes of practice, technologies, and scientific breakthroughs due to the rising rate of scientific and technological innovation. Engineering has increased man's quality of life through the creation of new and improved products as an applied science. There is a need to improve the economic efficiency and long-term viability of indoor fish farming. The need for and urgency of sustainable aquaculture development has passed the stage of planning. The future of farming is sustainable indoor fish aquaculture (Figure 1.1) (Blidariu & Grozea, 2011). Increased productivity with less environmental effect, system integration, and chemical reduction are only a few of the leading concepts that more sustainable fish production must follow. On a global scale, food safety for human use is becoming increasingly crucial. Aquaculture is a type of farming in which commercial fish are raised in containers, ponds, or tanks.

**Figure 1.1.** Indoor fish aquaculture.

*Source: https://upload.wikimedia.org/wikipedia/commons/2/2c/Shrimp_hatchery.jpg.*

A recirculating aquaculture system (RAS) is an aquaculture system that integrates water treatment and reuse while replacing less than 10% of total water volume per day. The idea behind RAS is to reuse a volume of water by treating it and delivering it to the organisms being cultured. RAS water treatment components must be able to handle the large amounts of feed required to maintain high rates of growth and stocking densities required to satisfy the financial goals. A RAS typically includes mechanical and biological filtration components, pumps, and storage tanks, as well as a variety of additional water treatment elements that increase water quality and disease management.

**Figure 1.2.** Hydroponics.

*Source: https://iberflora.feriavalencia.com/wp-content/uploads/2018/04/CDC_South_Aquaponics_Raft_Tank_1_2010–07–17–1024x768.jpg.*

## 1.2. HYDROPONICS

Plant roots develop in nutritional solutions with or without mechanical assistance from an artificial media. Hydroponics is a type of plant culture that allows plants to grow in nutrient solutions with the help of inert substrata. Hydroponics (Figure 1.2) is a promising approach not only for plant physiology investigations but also for commercial crop production (Jensen, 1999). It is widely used in both laboratory experiments and commercial crop production. Both aquaculture and hydroponics have drawbacks. Hydroponics necessitates the use of expensive nutrients to feed the plants, as well as periodic system cleaning, which can cause waste disposal concerns. Excess nutrients in recirculating aquaculture must be eliminated from the system, which usually involves removing a percentage of the water on a daily basis. After that, the nutrient-rich water must be discarded and replaced with clean, fresh water. While both recirculating aquaculture and hydroponics are highly effective means of producing fish and vegetables, when the two are combined, these disadvantages become advantages. The most prevalent water quality issues in RASs are oxygen depletion and organic matter accumulation, as well as inorganic nitrogen, mainly ammonia, and CO. These issues have historically been addressed by water exchange. In a fish production system, feed is continuously delivered, but effluent accumulates in dangerous amounts. Because it mostly consists of feces, uneaten food, and bacterial biomass, wastewater discharge is high in nutrients. Ammonia is produced as food decomposes. Fish emit ammonia at the gills during the metabolic processes of their growth. Large volumes of water dilute the ammonia in case of lakes and oceans. It must be controlled while keeping fish in containers, ponds, or tanks because it is extremely hazardous to them. The elimination of ammonia, a metabolic waste product excreted through the gills of fish, is a key challenge in aquaponic systems. Unless ammonia is eliminated through the process of nitrification, in which ammonia is oxidized first to nitrite, which is toxic, and then to nitrate, which is comparatively non-toxic, ammonia will accumulate and reach toxic levels. This two-step process is mediated by two types of naturally occurring bacteria like *Nitrosomonas sp.* and *Nitrobacter sp.* Feed leaching and bacterial breakdown of fecal contents generate total dissolved solids (TDS). The solid separation unit separates total suspended solids (TSS), which includes fecal solids and bacteria biomass, from water, and then discharges it as sludge from the culture system.

Excessive ammonia has a number of negative consequences, including extensive tissue damage, particularly to the gills and kidneys (Wilkie, 1997). We know that nitrite is less toxic to fish than ammonia, but that doesn't mean it's a good thing. The nitrite binds to hemoglobin, resulting in methemoglobin, which is unable to transport oxygen. With *Nitrobacter sp.*, nature once again assists us. These beneficial bacteria take nitrite and convert it to nitrate as part of their metabolic process. Nitrates are not a major issue for fish, but we must manage them at a low level to ensure their health. This is advantageous for humans because nitrate is the preferred food of plants. Furthermore, nitrate is far more tolerable to fish than ammonia or nitrite. Flow-through or recirculating intensive fish culture systems can be found suitable for fish production. Clean water is normally sent through the production unit in a single pass before being discharged or dispersed. Freshwater flow-through or open systems are prevalent in areas where clean water is abundant. Recirculating systems, on the other hand, provide capabilities for maintaining acceptable water quality. These systems have the potential to be more environmentally friendly because water input is reduced and discharged water is treated. However, new environmental regulation trends are limiting the amount of water that can be consumed or discharged, limiting the ability to use large influxes of water to remediate excess nitrate. This problem can be solved by using vegetable crops as a biofilter for aquaculture wastewater (Gregory et al., 2012). If fish have a low business cycle, we have our plant revenue to fall back on, and vice versa. Production crop capabilities are widely estimated. Some industry experts believe it has the ability to produce more than traditional or hydroponic farming, while others believe it produces far less. Fish and plant integration is a type of polyculture that increases diversity and consequently improves system stability. The sale of greenhouse products creates revenue that helps to sustain the local economy. Aquaponics improves economic efficiency by lowering numerous critical costs including nutrients, land, water and cost of infrastructure.

Aquaponics is a recirculating culture system that combines fish with hydroponic plants. The system was built with widely available materials and was designed to reduce capital expenditures, energy use, water consumption as well as the technological skills required to operate it (Enduta, Jusoh, Ali, & Wan Nik, 2011). Ammonia, which is hazardous to fish in its non-ionized form, is one of these waste products.

Ammonia is transformed to nitrite, which is then oxidized to nitrate, and biofilters are frequently used to aid in the digestion of these waste products. Chemical filtration is used to control the accumulation of end products (Danish et al., 2021). Partial cleaning is commonly used in big aquariums or fish culture systems where other techniques of reconditioning are expensive. However, this process consumes a lot of water and may contribute to pollution. Fish and plants are grown in such systems in a mutually beneficial, symbiotic connection.

The fish produce unionized ammonia-nitrogen as an intermediate by-product of protein synthesis, and increased levels of this nitrogen can cause death. Some forms of nitrogen, on the other hand, can be used as a plant nutrient and are extracted from the water when the water circulates through the hydroponics unit by the plant roots. As a result, a potentially dangerous by-product of fish production becomes a valuable plant fertilizer (Olsen & Hasan, 2012).

## 1.3. EFFLUENTS

Nutrients in aquaculture effluent are used to grow plant crops. The effluent is treated in the plant component before being reintroduced to the fish culture tanks (Cho, Hynes, Wood, & Yoshida, 1994). Aquaculture effluent normally provides enough amounts of 10 of the 13 needed plant nutrients, with only Ca, K, and Fe requiring supplementation. Continuous nutrient synthesis from fish waste prevents nutrient depletion, whereas plant uptake of nutrients prevents nutrient buildup. These increases water reuse and reduces discharge to the environment. Culture water can be reused for several years. Water quality monitoring is also reduced due to the balance between nutrient formation and removal. The acid created by the nitrification of ammonia, which is expelled by fish through their gills (Figure 1.3) causes pH to drop on a regular basis.

**Figure 1.3.** Ammonia poisoning.

*Source: https://dlgdxii3fgupk.cloudfront.net/myaquariumclub.com/images/fbfiles/images/image-bec29e6a94301e0433d501ad4b21a34d_v_1421890205.jpg.*

Maintaining an optimal ratio between daily fish feed input and plant growing area achieves nutrient balance. Basil yields were equal in batch and staggered production, however the growth of all plants in the same phase resulted in nutrient depletion in the culture water and the beginning of nutritional deficiency disease in the basil. When four development phases of basil were staggered, no deficiency symptoms occurred, despite the feeding ratio being higher due to elevated water temperatures and a good fish feeding response during the batch production trial. The high nutritional demand of plants in the end growth phases was counterbalanced by lower nutrient demand of plants in the early growth stages, resulting in nutrient intake being moderated and nutrient depletion avoided. More research could be done to discover the best feeding ratio for staggered basil production and whether batch culture of basil is possible at a greater feeding ratio. It is suggested that a staggered production approach be utilized in aquaponic systems, especially for crops with short production cycles like lettuce and basil.

**Figure 1.4.** Aquaponics.

*Source: https://www.kodamakoifarm.com/wp-content/uploads/2020/08/koi-aquaponics-preview.jpg.*

Aquaponics (Figure 1.4) is not a new concept in many ways, but interest in raising fish and plants in an integrated, indoor system has increased in recent years. According to USDA-NASS (2013), 21 states reported at least one aquaponics farm in 2012, for a total of 71 aquaponics farms in the United States. Aquaponics farms made up 2% of all aquaculture operations (Zappernick, 2020). 75% of these had sales of less than $25,000, compared to only 48% of pond-based operations. Another 14% made sales ranging from $25,000 to $49,000, and 11% made sales of $50,000 or more. In comparison, 60% of pond-based businesses made $50,000 or more in sales. Florida had the most aquaponics farms (20%) in 2012, followed by Wisconsin (10%),

Arizona (8%), New York (8%), and Hawaii (8%). Hawaii had the largest aquaponics farm (4,741 gallons), followed by Arizona (3,208 gallons), and Wisconsin (3,208 gallons) (2,004 gallons) (Froehlich et al., 2021). With an average capacity of 537 gallons, the aquaponics farms in Florida were substantially smaller. The tiny scale of aquaponics farms suggests that they are mostly run as a hobby or lifestyle choice, with some extra income, rather than as full-time aquaculture operations (Love et al., 2015). This distinction is critical when discussing aquaponics economics. Individuals who aim to use aquaponics as a form of home gardening activity will not need to be as concerned with costs and revenue as those who plan to use aquaponics to support their family. Despite the diversity, three broad types of systems exist: raft or deep-water culture systems, nutrient film systems, and media-filled bed systems.

**Figure 1.5.** Raft culture.

*Source: https://www.researchgate.net/publication/275213656/Figure/fig3/AS:667642834739208@1536189864164/The-raft-system-for-mussel-culture_W640.jpg.*

Raft culture (Figure 1.5) is often favored for commercial operations, but the nutrient film method (NFT) employed in hydroponics is limited to plants with small, light root systems such as leafy green vegetables. Solids must be eliminated in both raft and NFT systems. Media-filled beds are more widely utilized for home-based aquaponics cultivation than raft systems, and therefore require lower stocking rates.

## 1.4. INVESTMENT

The overall investment required to construct facilities and purchase necessary equipment, the annual costs to operate the system and realistic estimates of market prices, the degree of competition in the markets to be targeted, and realistic estimates of revenue to be received are all important economic considerations for any type of business. Estimating the amount of money needed to build an aquaponics system is perhaps the easiest stage. Costs for various types of tanks, PVC, pumps, and filters are widely available from supply stores, as are cost estimates for a greenhouse. All of the necessary components of the business must be carefully planned. Given that many of these technologies are very new and have limited detailed assessments of their costs and returns over time, estimating annual operating expenses becomes a bit more complex. Particularly for the pounds of fish that may be raised, the volume of vegetables that can be produced, and the hazards associated, very conservative estimations must be employed. Power failures in the winter, for example, might result in the complete loss of a tilapia crop. Infestations of diseases or parasites can be difficult to control in aquaponics systems since pesticides can destroy the other crops that are part of the system. It's crucial to slightly underestimate the development and productivity of fish and plants while slightly overestimating the costs. A company plan that takes such a cautious approach is more likely to succeed. Developing a realistic, accurate, and workable marketing plan is the most difficult component of running an aquaponics firm. Fish are two to three times more expensive to raise indoors than in open ponds. As a result, a viable aquaponics enterprise will need to find and build a market for the produce that will pay a higher-than-average price. For example, an aquaponics lettuce grower will face competition from lettuce sold at Wal-Mart, other grocery shops, and farmers' markets.

Why would someone choose aquaponically farmed lettuce over other types, especially if it is more expensive? To be competitive, the seller must have a clear answer to that question. A second marketing concern is that a high-end market that is willing to pay a premium price will also incur higher marketing costs. If a top chef is paying a premium price for aquaponically farmed herbs because of the freshness of the produce, that chef may desire frequent deliveries to ensure that freshness. Frequent delivery will necessitate additional staff, cars, and mileage costs. It's also important to think about the labor requirements. An aquaponics system necessitates regular monitoring. Aquaponics systems, even on a small size, are complicated due to their numerous components and requirements. Harvesting and packing vegetables takes a lot of time and effort. Labor expenditures accounted for 46% of total operational costs and 40% of total annual costs.

Aquaponics is a closed-loop production method that combines RASs and hydroponics to produce fish and plants in a way that mimics nature's ecology (Masabni & Niu, 2022). The fish excrete nutrient-rich waste that fertilizes the plants, while the plants filter the water for the fish. Bacteria break down fish waste from the aquaculture portion of the system into dissolved nutrients, which plants use to grow in the hydroponic component. This nitrogen removal not only improves fish water quality, but it also reduces total water usage by minimizing the amount of effluent emitted. The general public's idea of aquaponics' sustainability is based on the synergistic link between fish and plants. Aquaponics can also be scaled up or down, from a small hobby unit to multi-acre commercial production plants. Producing high-value herbs, vegetables, as well as fish, crayfish, worms, and a variety of other items, allows the producer to cater to a diverse clientele. Because aquaponics is a closed-loop system, nutrient contamination is almost non-existent, allowing agriculture to make a significant contribution to environmental sustainability. Furthermore, fish, plants, and waste materials can all be gathered and transformed into fertilizer for resale. These advantages make aquaponics a realistic choice for gardeners and producers with limited space or who live in cities, increasing access to locally produced, healthy foods for more people.

**Figure 1.6.** Recirculating aquaculture systems.

*Source: https://upload.wikimedia.org/wikipedia/commons/9/95/Recirculating_Aquaculture_System_7.jpg.*

## 1.5. RECIRCULATING AQUACULTURE SYSTEM

An aquaponics system must include several essential components in order to function properly. RASs (Figure 1.6) are self-contained aquatic production habitats (Martins et al., 2010). RAS can vary widely in terms of price, components, and production capacities, but all systems must include a location for particle removal, biological filtration, water flow, and the culture animals to dwell. In general, the size of the fish and the size of the tank are proportional. The fish dwell in the culture tank as they grow to the desired market size. The fish's needs for space, food, water quality, illumination, flow, and other elements will alter as they grow. In this area, it is critical to maintain acceptable water quality, proper oxygen levels, optimal flow rates, and waste removal. Aeration agitates the water, resulting in a high surface area for air and water contact, which improves gas exchange. The more the gas exchanges, the smaller the bubbles must be. To facilitate the elimination of carbon dioxide and enhance the concentration of dissolved oxygen, aeration should be given in the culture tank. Air bubbles can help some fish species feel more hidden and at ease, as well as improve eating behavior. It's also vital to provide a cover to keep fish from jumping out of the tanks; some species, such as tilapia, are strong jumpers that will jump out and die. It's best to use a strong lid made of waterproof material that completely covers the tank. However, because gas exchange is vital for fish respiration, small openings must be left to keep the fish in while allowing air to flow in and out. Plywood, foam insulation, or plastic mesh attached to a polyvinyl chloride (PVC) frame can be used to cover the tank (Mohapatra, Sarkar, Barik, & P, 2011). A mesh cover provides for simple feeding and gas exchange, but an opaque cover reduces sunlight penetration into the tanks, lowering stress on the fish, algae growth, and temperature fluctuations caused by solar inputs.

To achieve rapid development rates in RAS, a comprehensive food with suitable protein and fat levels is required. Unfortunately, fish cannot fully utilize the feed; roughly 25–50% of the feed will become waste, which must be removed from the system as soon as possible to minimize water quality degradation (Barlow, 2003). The size of these waste materials ranges from less than 10 microns to more than 100 microns. If the particles aren't removed, bacteria will break down the feed, resulting in an oxygen shortage, which might kill the fish and beneficial bacteria in the biofilter. It can also result in the emission of ammonia, which is hazardous to fish. To remove these materials, filters have been designed that can remove the solids and perform the functions of a biofilter, improving the RAS's operational efficiency. Mechanical filters include low-cost filter pads that must be cleaned by hand, passive water clarifiers, fluidized sand and bead filters that combine solids and biological filtration, and self-cleaning microscreen drum filters with huge volumes and costs (Hasan & Andriani, 2018). Filter pads, such as those found in industrial air handling units are a low-cost option, but they must be cleaned on a daily basis. Solids must be cleaned from the filter more frequently at increased fish capacity and feeding rates, and automated filters may be required to decrease labor expenses.

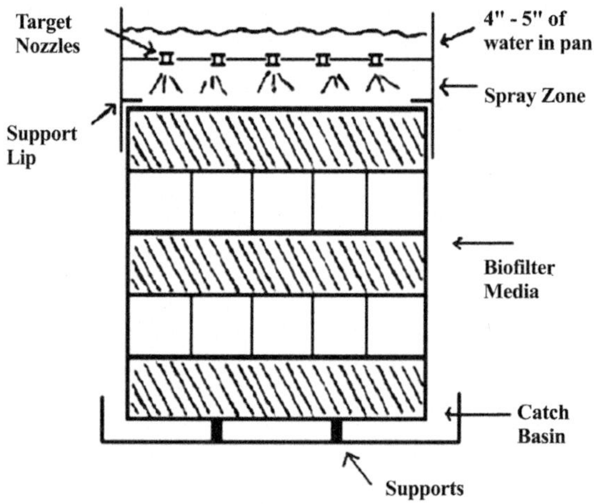


**Figure 1.7.** Biological filter.

*Source: http://biofilters.com/webfiltjan2015_files/image004.gif.*

## 1.6. BIOLOGICAL FILTER

The biological filter (Figure 1.7) uses bacterial processes to transform toxic by-products of fish feeding into non-toxic forms that plants can use. The biofilter is made up of a substrate on which beneficial bacteria can grow in high densities with plenty of water and oxygen to convert toxic ammonia and nitrite to nitrate (Bartelme, McLellan, & Newton, 2017). Surface area for bacterial attachment, temperature, dissolved oxygen, alkalinity, pH, availability of ammonia and nitrite, and other factors all limit the biological breakdown process. To maximize toxin conversion, most biological filters use a high surface area medium that is resistant to breakdown and allows space for water flow and gas transfer. In addition to toxin conversion, some biological filters can also act as solids filtering or degassing chambers, which improves operational efficiency. There are a variety of biofilter designs and substrate materials to choose from, and some mechanical filters can also serve as biofilters. The least expensive and lightest weight product that allows water and air to pass through without clogging is usually the best choice. PVC ribbon, which is inexpensive, light weight, and has a large surface area, can be an effective biofilter solution (Soliman, Morad, Wasfy, & Moursy, 2020). The sump tank, which is located at the lowest elevation in the entire system, is the ultimate collection area for water that flows by gravity. The sump tank is also a reservoir with no plants or fish, and it's a good area to mix chemical and nutrient additives that need to dissolve before entering the system. If needed, add more biofilter substrate or cleaner fish like mosquito fish, redear sunfish, or juvenile tilapia to the sump tank, but keep them away from the pump input.

Pumps are utilized to move water between the various components of the system. In order to improve efficiency, there are several methods for transporting water and many engineering concerns that should be integrated into system design (Parker & Suttle, 1987). When sizing a pump, the flow dynamics of water should be taken into account. Water flow velocity, for example, is vital for keeping pipes clean on the inside, and the head pressure influences the total flow rate that a pump can produce. To create water flow, impeller pumps use the centrifugal force of an impeller within a chamber with a single inlet and single outlet. Impeller pumps are typically powered by electricity; however, some versions can also be powered by fuel combustion if energy is unavailable (Mahmudov, Mahmoud, Sur, Cruz, & Bilton, 2019). Because these pumps are powered by electricity, the electrical components must be waterproof. Impeller pumps are available in submersible and in-line versions, but they all work on the same concept. Submersible pumps' electrical components are enclosed in a plastic or insulated metal casing. Submersible pumps have the advantage of always having access to the water they pump, unless the sump tank goes dry. Inline pumps have a longer life expectancy and greater volume and pressure rating possibilities than submersible pumps (Davidson, Helwig, & Summerfelt, 2008). In-line pumps, on the other hand, must be primed before moving water, and depending on how they're installed, losing supply, especially during power outages, can be a serious problem. Airlift pumps use air bubbles to transport water. A simple airlift pump will pump air into a pipe that is submerged in water, forcing water through the pipe as the bubbles rise. The combined action of aeration and water movement makes airlift pumps effective. The air produced by the blower can also be used to provide aeration to the deep-water culture unit, which is another advantage of using an airlift system. Airlift pumps are good for providing directional flow, but their pumping height is restricted, therefore they may not be suitable for many applications (Parker & Suttle, 1987). However, advances in airlift technology may enable them to transfer water vertically up to 20 feet. Plumbing is used to move water between system components and comes in a variety of sizes and materials, depending on the use (Costa-Pierce, 1987). Because it is robust, light weight, relatively inexpensive, easy to work with, and generally available, PVC is commonly used in RASs. Vinyl tubing can be employed in some applications that aren't subjected to a lot of pressure. Steel is the strongest pipe material and may be ideal for some purposes; but, when exposed to salt water, steel is particularly prone to rust and corrosion. Because of its inherent antimicrobial characteristics and capacity to withstand extremely high-water temperatures, copper is often employed in household applications. Copper is costly, requires skill to connect, and can be poisonous to invertebrates such as crayfish, shrimp, and prawns. The pipe diameter should be calculated based on the volume, pressure, temperature, and head loss of the water that will be encountered in various parts of the system. Pressurized water pipes, such as the inflow portion to the culture tank and some solids filters, should always be glued together. To guarantee that non-pressurized drain lines can manage the flow rate required without clogging and creating overflows, they should be sized bigger than pressurized drain lines. The pipe diameter should be chosen to prevent biofouling while maintaining the necessary flow rates to the tanks.

In recirculating aquaculture, gas exchange is the act of adding or removing chemical components that are dissolved in the water (Loyless & Malone, 1998). In recirculating aquaculture, oxygen is essential for biological activities in both fish and beneficial microbes. Nitrogen digestion and nutrient uptake in plant roots are aided by high dissolved oxygen levels. In the fish tank and hydroponic unit, air is normally injected as minute bubbles via air stones. When possible, free aeration should be used, such as a waterfall from a pipe into a reservoir, trickling water over filter pads, and biofilter substrate. Gas exchange can also be used to remove potentially harmful gases such as carbon dioxide, nitrogen gas, hydrogen sulfide, and others. Aeration is the process of bringing a large amount of air into contact with water. This is usually accomplished by dropping water into the air or infusing air into the water through a diffuser or porous media. In this process, the more surface area exposed between the air and the water, the greater gas exchange. Keep in mind that water reacts to the air it is exposed to, so if oxygen levels are high in the air, they will be high in the water. The same is true for carbon dioxide, which will tend to build up in buildings with poor ventilation. Carbon dioxide is a heavy gas that tends to settle close to the ground. When venting carbon dioxide from a RAS, more attention should be exercised. Injecting air into the water can be done in a variety of ways. A blower is suited for high-volume, low-pressure applications. Blowers are limited by water depth, and they work best in less than three feet. Compressors overcome the depth restriction by producing high pressure air, however their volume is limited (Barrut, Blancheton, Champagne, & Grasmick, 2012). At depths more than eight feet, compressor systems are often employed in ponds for vertical mixing of the water rather than gas exchange. A venturi is an air injection device that uses the vacuum formed by pressured water movement to inject air. A venturi nozzle, in essence, suctions air into the water flow stream, resulting in small bubbles and efficient gas exchange. The potential for supersaturation of a gas under high pressure in the water, which can come out of suspension in the fish's blood, leading to gas bubble disease, is one concern with venturi systems.

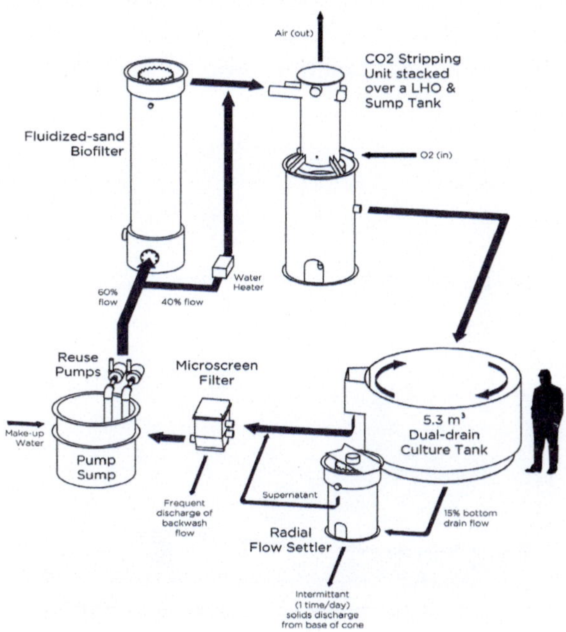


**Figure 1.8.** Ozonation.

*Source: https://www.hatcheryinternational.com/wp-content/uploads/2019/12/6-RAS-PFD_11.2019–550x625.jpg.*

For gas exchange, water can be sprayed or dripped into the air. To increase gas exchange, a propeller or agitator-style unit creates a disruption at the air-water interface. Fish must be kept out of this area since the unit's blades can harm them. A degassing tower creates tiny water droplets by creating a controlled waterfall effect over a hard and porous substrate. These are quite effective and affordable. These approaches do result in a significant amount of evaporation and humidity in the air, which can lead to corrosion and the breakdown of equipment or structures, as well as the possibility of mold. This keeps the water clear and still while also delivering a lot of oxygen, which helps with solids removal. The possibility for supersaturation and the genesis of gas bubble disease in fish is a serious concern with pure oxygen infusion. Furthermore, pure oxygen can be rather costly.

Ozonation (Figure 1.8) is the technique of creating ozone by putting an electric current in contact with pure oxygen (Wietz, Hall, & Høj, 2009). Ozonation can kill microbial pathogens in the water, and it provides high dissolved oxygen levels in the water when it breaks down from $O_3$ to $O_2$. Ozone, on the other hand, poses a risk to fish, beneficial bacteria, humans, and the equipment itself due to its high oxidative power. Working with professionals to establish an ozonation process with a high level of safety for the system is crucial.

**Figure 1.9.** Aerator.

*Source: https://www.aireo2.com/wp-content/uploads/2015/06/Aquafinca-2014–06–03–009VEN-e1480975354247.jpg.*

An aerator (Figure 1.9) a blower, or a compressor can be used to add air to the water. Aerators are affordable; however, they have a restricted volume. Blowers are more efficient, but they can't force air more than three feet down into a water column because they don't produce enough pressure. To force air deeper than three feet, a compressor is required, but it is more expensive to operate and does not generate as much air volume. The management procedures appropriate for purifying the water prior to entering the system will be determined by the water source used like groundwater, surface runoff, rainwater collected, or municipal supply. Water from municipal sources may be cold, have a high pH, and include chlorine, including chloramines. Chloramine is an odorless molecule that is exceedingly stable in water and requires the use of sodium sulfite or sodium thiosulfate to break it down into its constituents of chlorine and ammonia (Gaikowski, Larson, Steuer, & Gingerich, 2004). Chlorine and ammonia are both hazardous to fish and must be treated before being added to the system. Lowered DO and pH levels, as well as the release of ammonia from the dichlorination process, are concerns related to the use of sodium sulfite or thiosulfate; thus, water quality must be assessed prior to introduction into the aquaponic system.

## 1.7. TEMPERATURE

Temperature has a direct relationship with fish growth and beneficial bacterial efficiency. Each species can endure a wide variety of temperatures, with an ideal temperature that should be maintained for maximum development rates (Pankhurst & King, 2010). To obtain this ideal temperature, the water may need to be heated and cooled. It's critical to consider the energy costs of raising or lowering water temperatures, particularly in light of the surrounding environment. In a well-insulated building that does not get solar inputs, temperature management is much easier. Temperature variations throughout the year, especially in temperate areas, can make a significant difference in a company's profitability. To achieve optimal effectiveness, the fish species should be matched to the current temperatures. Heat is exchanged by surface-to-water contact in heaters and coolers. Temperature regulation will be faster and more efficient if the air temperature is adjusted to meet the target water temperature. Furthermore, while renewable energy sources such as geothermal, solar, and wind may be sustainable, they might have significant installation and running costs that are mitigated by some government programs. A counter current exchanger is used in boiler systems to transfer heat from one source to another. To improve sustainability, the heating source might be woody biomass or other waste items.

Pathogens pose a threat to the health of plants, fish, and humans. Cross contamination of the product is possible when fish excrement gets into close touch with plant roots (Defoirdt, 2014). GAPs or good agricultural practices can help to limit the danger of contamination. Clean, pathogen-free water is good for growing fish at their maximum rate (Amekawa, 2009). Diseases can spread quickly in recirculation systems due to the constant reuse of water and close quarters of the fish produced. Due to the fact that infected fish do not eat, there is a substantial risk of water quality degradation, especially in highly automated systems. Aquaculture operations involve ultraviolet (UV) irradiation and ozonation. Irradiation with UV light is an excellent way to kill single-celled bacteria in the water column. An UV light sterilizer is a device that reduces pathogen levels in water. UV sterilizers are frequently used in RAS systems in conjunction with ozone to eliminate microorganisms (Gullian, Espinosa-Faller, Núñez, & López-Barahona, 2012). To be successful, UV must be exposed to bacteria for a particular amount of time, hence flow velocity is critical. For larger species or when the water becomes dirty, UV is ineffective. UV bulbs also degrade over time and should be replaced on a regular basis. Ozonation is the technique of producing ozone by putting an electric current in contact with pure oxygen. Ozone is a highly reactive oxidizing gas that kills organic material, including bacteria and viruses. Ozone can damage fish, useful bacteria, humans, and equipment due to its oxidative power. Ozone is especially problematic in saltwater systems because it combines with bromide in the water, potentially producing poisons.

## 1.8. MONITORING

Environmental monitoring is crucial to recirculating aquaculture's success. Facility managers can do this manually or remotely via monitors. Knowing the state of a system in terms of energy, temperature, water quality, humidity, and security may be considered a luxury for small businesses, but it is a vital risk management activity for larger enterprises (Sánchez-Mata & Mora, 2000). Many automated monitoring systems include an alarm system that notifies facility managers of potential dangers, and some even allow managers to control the growing environment from their smartphone or computer. A backup generator that kicks in automatically in the event of a power loss will also rescue fish, but a backup of automatic oxygen diffusion into the tanks during an outage should also be included in the contingency plan. When rearing fish at high densities, it is necessary to continuously check and regulate water quality. Humans, unlike other livestock producers, have a hard time detecting when environmental conditions are outside of their ideal range before it's too late. At least once a day, dissolved oxygen, pH, temperature, ammonia, and nitrite are monitored, and more frequently if an issue is detected. Carbon dioxide and ammonia levels rise as fish feed, oxygen, and pH levels decline. To maintain fish survival, intensive water quality management is required to minimize these changes. There is a wide range of testing equipment available at various prices.

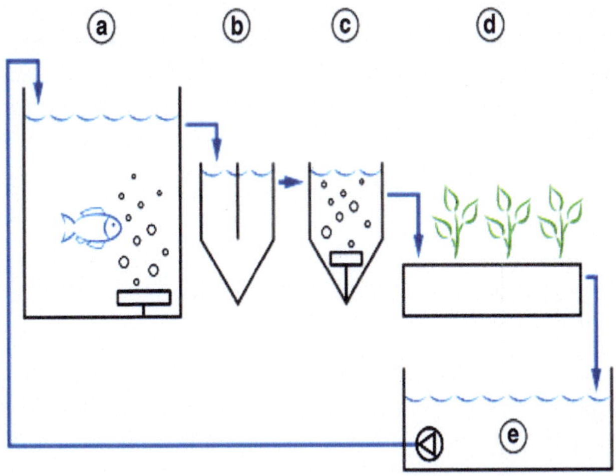


**Figure 1.10. Coupled aquaponics system.**

*Source: https://www.researchgate.net/publication/350416152/Figure/fig1/AS:1005840228159488@1616822406290/Schematic-of-a-basic-coupled-aquaponics-system-with-a-fish-tank-b-sedimenter-or_W640.jpg.*

When selecting a species, keep in mind the organism's temperature, salinity, and protein requirements, as these factors influence growing expenses as well as the species' ability to be integrated into integrated systems like aquaponics. Be aware that any specific species may have regulatory complexity, and that appropriate regulatory agencies should be consulted to confirm that all necessary permissions are in place before importing items. In terms of a starting facility and risk management, the ease or difficulty of growing that organism should also be considered. Some species are far too sensitive to water quality variations to be deemed suitable for cultivation. When forming a business model around an individual species, the value of the product and markets must be addressed in addition to all of the other aforementioned variables. In general, the less system complexity required for successful production of a given species grows as the ease of culture increases from low to high. Coupled and de-coupled aquaponics systems are the two basic types of aquaponics systems. Coupled, or balanced, systems work on the assumption that the fish's incoming feed meets the exact nutrient needs of the plants being cultivated. In theory, nutrients, such as nitrate, are scraped from the water each time it goes through the plant culture unit, and feed additives are required to meet further plant needs. The rate of nutrient uptake is affected by factors such as feeding time, feed volume, fish, and plant size, temperature, pH, and water chemistry.

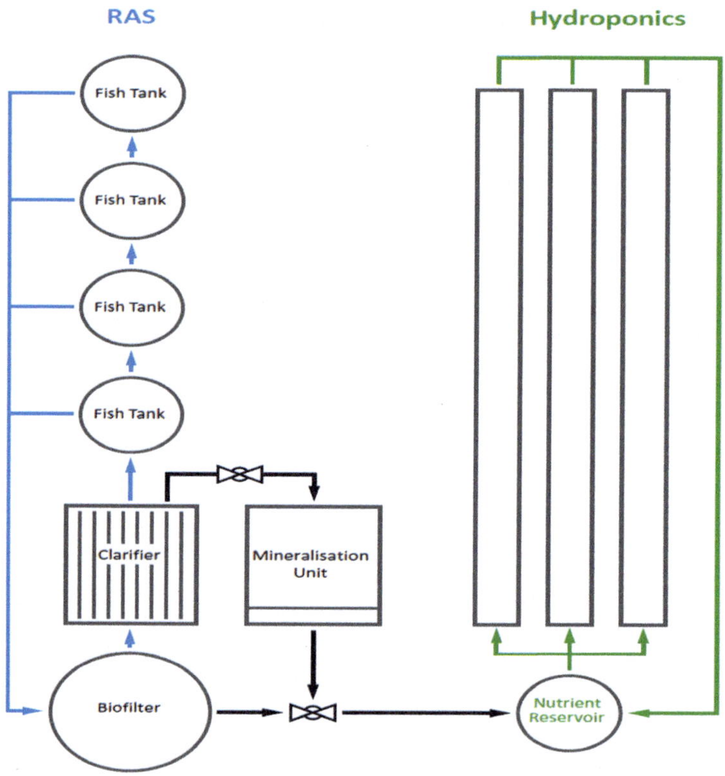


**Figure 1.11.** De-coupled aquaponics system.

*Source: https://www.researchgate.net/profile/Simon-Goddek/publication/333938727/Figure/fig3/AS:773238730752O4@1561386187727/Schematic-picture-of-a-decoupled-aquaponic-system-including-an-aerobic-mineralization_W640.jpg.*

Water in a coupled aquaponics system (Figure 1.10) usually flows in one direction, starting in the fish culture unit and going through a solids filter, biological filter, hydroponic unit, and sump tank before returning to the fish and completing the cycle (Baganz et al., 2022). Incoming feed uniformity, solids, and biological filtration, and plant nutrient uptake are all crucial to the consistent and predictable generation of products to market. Producers gain greatly since fish feed typically provides 10 of the 13 macro and micronutrients required for plant growth. Calcium, iron, and potassium are three nutrients that are commonly deficient and may require supplementation (Rasdi & Qin, 2016). Because some ionic nutrients, such as phosphorus, are present in feed, a mineralization or digesting stage is required to break it down and free all of the nutrients. Unless chelated by a non-ionic binder like Ethylenediaminetetraacetic acid, iron will precipitate out of solution as iron oxide. In a balanced system, iron is usually given at a rate of 2 mg/L every two weeks. It's practically impossible to treat one of these components without influencing the others because of the delicate balance of fish, bacteria, and plants. In RAS, for example, using salt to treat fish disease is a frequent and safe technique, but in aquaponics, salt, especially chloride, is a major disruptor of plant growth. In contrast, at high enough concentrations, organic certified insecticides like neem oil can harm fish.

De-coupled aquaponics systems (Figure 1.11) separate the fish and plant sections of the overall aquaponics cycle, allowing for disease or catastrophic loss in one portion of the system to be treated without causing the entire system to collapse (Blanchard, 2019). A typical de-coupled system includes a stand-alone RAS system with effluent water that drains into a collection area and is used in the independent hydroponic system. Because the hydroponic unit has several water sources, it may operate with inputs from the RAS effluent or a municipal supply if there are any difficulties with salt or other fish drugs. A de-coupled system helps to eliminate the trade-offs in pH, temperature, and other water quality factors. For example, nitrifying bacteria require a slightly alkaline pH of 7.0–8.5 to convert ammonia to nitrate, whereas plants absorb nutrients best around pH 5.0–6.5. pH can be adjusted in a de-coupled aquaponics system to maximize nitrification and plant development (Kuhn, Drahos, Marsh, & Flick, 2010). Furthermore, the temperature at which the fish are cultured no longer forces the producer to grow a certain plant species for that temperature range. If local market conditions indicate that a cold-water fish such as trout is best, but the plant market prefers warm-weather herbs such as basil, both species can be optimized at the same time without causing harm to one another. Finally, facility managers can adjust the water as needed if additional nutrients are required for a specific plant variety.

Aquaponics can be done on a large scale or small scale, ranging from a hobbyist's bench-top aquarium to a multi acre commercial operation capable of producing large quantities of fish and plants every year. Profitability in the aquaponics business model is linked to scale and efficiency of production, just like in other agricultural businesses. Cultivating tilapia and basil in a greenhouse facility in Iowa could be profitable. This system model shows that the value of the fish produced has very little impact on profitability, and that the price and quantity of plants produced are often the determining factors. Aquaponics may be an appealing option for people who desire to live a more relaxed lifestyle while earning a modest income. Because of improved product output, efficient resource utilization, system stability, and regularity of production, the ability to profit grows as the plant growing area grows in a well-designed and efficiently maintained aquaponics facility. A larger facility, on the other hand, does not always imply higher profit. To assess the actual returns to the farmer, one must consider supply and demand fundamentals as well as wholesale versus retail pricing. To assure success, potential aquaponic farmers must complete their due diligence in business planning and market research, as well as hands-on education, before investing in an aquaponics firm.

The basic idea is to use water efficiently to grow two crops rather than one, and to divide and share nutrient resources between fish and plants. This farming approach is often used to cultivate both fish and vegetables in an integrated system in resource-constrained and urban settings. Aquaponics combines the cultivation of horticulture plants with the raising of fish. Many plants are suited for aquaponics systems, but which will function best for a given system is determined by the fish's maturity and stocking density. Green leafy vegetables with low to medium nutrient requirements, such as capsicum, tomatoes, lettuce, cabbage, lettuce, basil, spinach, etc., are well adapted to aquaponics systems (Figure 1.12) (Wuang, Khin, Chua, & Luo, 2016). It's essentially a Recirculation Culture System in which fish are given high-quality floating pellet feed and waste is pushed into bio-filter troughs. When using horticultural plants, the water flow rate must be controlled using a timer. The aquaponics system's fish and plants are completely organic. Even though the system's initial investment is large, the system's ongoing costs are low, and it provides reasonable returns. This approach has the advantages of consuming less water, less land, waste renewal, and less labor, among other things. Plants and animals in an Aquaponic System work together in a symbiotic relationship. Aquaponic systems are believed to use only 2 to 10% of the water used in typical vegetable or crop production while producing 10 times the yield without the use of hazardous chemicals, pesticides, or other chemicals.

**Figure 1.12.** Vegetation and aquaponics system.

*Source: https://www.offthegridnews.com/wp-content/uploads/2014/09/aquaponics-2-aquaponicsplan.jpg.*

Aquaculture involves the growing of aquatic animals, while hydroponics involves the production of vegetable crops using nutritive solutions. Plants grown from aquaculture waste are an excellent advantage of aquaponics since they reduce contamination and boost water efficiency while also reducing environmental effect. Aquaponics can lower production costs by more efficiently using resources, can be set up at various production scales, boosts economic profitability, and diversifies financial income streams. In the aquaculture business, particularly in poor countries, excessive, and unregulated use of antibiotics, pesticides, and fertilizers is a global issue. Increased pressure on water resources and water quality could result in a drop in aquaculture production and have an impact on food security. Nonetheless, developments in aquaculture technology, techniques, and location have allowed for significant reductions in these environmental hazards and impacts. Traditional food production techniques, such as aquaponics, can address and reduce environmental pressures produced by aquaculture integrated systems. Aquaponics is the combination of two profitable and well-established food production technologies: aquaculture and hydroponics. In one integrated system, it merges and synergistically blends aquaculture and hydroponic processes. Due to short supply chains and organic fresh food, aquaponics could represent a new integrated agricultural system from producers to consumers. Aquaponics may also be a long-term solution to the restricted availability of fish and vegetables in some areas, because of its dual purpose. This technology has the potential to manufacture protein with less labor, land, chemicals, and water.

Human production of nitrogen and phosphorous compounds expanded rapidly, notably in the second half of the twentieth century, as a result of fertilizer use or as a by-product of fossil fuel combustion. Aquaculture effluents are high in nitrogen and phosphorus, making them a global concern for possible pollution. Aquaculture effluents are known to be high in dissolved and suspended solids, which are primarily made up of phosphate and nitrogen from fish excrement, feces, and uneaten feed. The growth of land-based aquaculture systems requires effective water filtration to remove waste material. Aquaponics systems were developed as a result of increased concentrations of these chemicals, which are also plant nutrients. The waste generated by the fish in the water tanks is directly utilized or transformed by bacteria into nutrients useful for plants in aquaponic systems.

Ammonia is the most common kind of nitrogen pollution in fish waste in the exit water. As a result, the water from the fish tanks should be directed first into the biological filters, where the Nitrosomonas bacteria may oxidize the ammonia to nitrite, and subsequently nitrite oxidizing bacteria, represented by the Nitrobacter, can convert the nitrite to nitrates (Johnson & Sieburth, 1976).

The water is then sent to plant growing units, where the nutrients are assimilated by the plants, and the water is cleaned before being recycled back to the fish tank. Because nitrogen is the most important nutrient for both plants and fish, the nutritional balance concentrates on it. If the biological filters fail to convert ammonia to the less hazardous nitrate, algae can be used as an alternative aquaponics vegetable crop, as it can use both nitrate and ammonia nitrogen.

Furthermore, aquaponics may maximize phosphorus usage to 71.7% of total phosphorus intake, indicating that aquaponics systems have the potential to become a new technique of recycling P and improving overall

P utilization. Emissions from aquaponic systems are mostly nitrogen and phosphorus discharge into the environment as a result of the removal and disposal of suspended solids and dead plants. These wastes, on the other hand, can be easily recycled within the farm because dead vegetables can be utilized to make humus by being mineralized by earthworms. Aquaponics, which combines aquaculture with agriculture, has the potential to improve nutrient efficiency and environmental sustainability.

# CHAPTER 2

# Recirculating Aquaculture

CONTENTS

## 2.1. INTRODUCTION

Over the last thirty years, recirculating aquaculture systems (RAS) have evolved through academic and commercial research, as well as ongoing improvement of each subsystem process (Ahmed & Turchini, 2021). Seafood can only be as delicious as the water it lives in. Aquaculturists monitor the water quality to ensure that the seafood they produce is free of toxins ("Publication : USDA ARS," n.d.). Cultured or farmed seafood has a strong consumer preference because it is more consistent in quality and appearance, and it has a milder flavor than wild seafood. Depending on the quantity of organisms cultivated per volume of water and the water source and supply, aquaculture systems can be extensive, semi-intensive, or intensive. RAS are intensive systems, pond culture is extensive while cage culture is semi-intensive system. Because pond and cage systems are open-air, pollutants from the air or water are always a possibility.

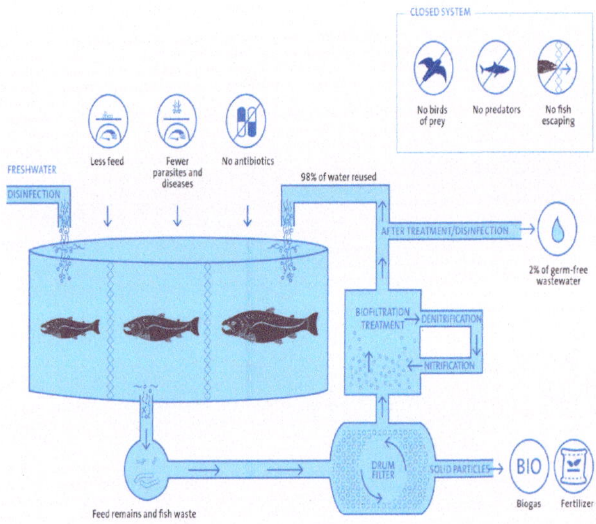


**Figure 2.1.** Indoor RAS.

*Source:https://journals.openedition.org/factsreports/docannexe/image/5766/img-2.png.*

Outdoor ponds and net pen-based systems have a substantial disadvantage in terms of disease risk, which could lead to loss of yield. Diseases in fish systems are spread through direct contact with infected organisms in the water. Indoor systems start with distilled water, and disease transmission is unlikely unless diseased or disease-carrying fish are put into the culture system. Outdoor pond and net pen-based systems are additionally hampered by their inability to offer a consistent product due to challenges in controlling growing cycles, resulting in supply peaks and troughs on the market (Gladue & Maxey, 1994). Problems with escapement are a big concern, especially with biotechnology modified species. Because the animals cannot leave an indoor RAS and hence have no impact on natural populations, RAS becomes the sole approved culture strategy in such instances.

## 2.2. INDOOR RAS

Indoor RAS (Figure 2.1) allows for controlled product growth rates and predictable harvesting schedules by raising fish in a controlled environment. RAS saves heat and water by reusing water after it has been reconditioned through biological filtration with biofilters (Van Rijn, 2013). RAS provides the most effective economies of scale of any aquaculture system, resulting in the highest production. RAS are environmentally friendly since they utilize 90–99% less water than traditional aquaculture systems, and take up less than 1% of the land area. Commercial RAS utilize less water, have better densities, and similar feed conversions (Song et al., 2019). RAS enables year-round production of predictable volumes of product as well as comprehensive environmental climate control. Aquaculture systems created by RAS are indefinitely scalable. Because wastes may be controlled in environmentally friendly ways, there are no environmental constraints to the size of the anticipated fish farm to be created. RAS allows for a great deal of environmental control. This not only reduces the hazards associated with outdoor aquaculture, but it also allows for year-round species growth. Indoor aquaculture is likely the only approach that may provide a completely clean source of seafood that is devoid of all toxins and heavy metals. With consumer doubts about food safety growing, aquaculture companies who use RAS have an once-in-a-lifetime opportunity to meet the need for healthy seafood. The benefits of RAS-produced seafood include fresher, safer, and locally raised goods. RAS systems have a competitive marketing advantage over outdoor tank and pond systems, which are seasonal and subject to environmental calamities beyond the control of the operator, because they can be set up to produce the same volume of fish every week (Asche, Roll, & Trollvik, 2009).

Even in Idaho, where NPDES permits limit fish farm effluents to monthly average suspended solids concentrations of 5 mg/L, serial-reuse systems may not need to capture large percentages of waste solids for their average discharge to meet concentration-based effluent limits on total suspended solids (TSS). Similarly, under typical fish production levels, such as 50 lb production per year per 1 gal/min or 6 kg/yr per Lpm, serial-reuse systems may not need to capture large percentages of waste solids because a simple mass balance shows that only about 5 mg/L of TSS concentration would be added to the flow if averaged over the entire day (MacMillan, Huddleston, Woolley, & Fothergill, 2003). Nutrient restrictions are also becoming a problem with discharge. However, because of the huge water volumes used in single-pass and serial-reuse production systems, removing nutrients from effluents is not feasible for farms like Clear Springs Trout Company (Everson et al., 2021). To reduce aquaculture's environmental impact, production strategies and technologies are being implemented to reduce waste output, conserve water, and concentrate wastes into smaller usable components during fish culture, which has sparked increased interest in RAS. Traditional flowing water systems can yield roughly 6 kg of fish per year for every 1 Lpm of water flow. Partial-reuse systems can produce as much as 48 kg of fish per year for every l Lpm of water flow, or 400 lb of fish per year for every gallon per minute of water flow, by reusing or recycling 80 to 90% of the water prior to release (P. Li et al., 2022).

## 2.3. AQUACULTURE

Aquaculture must continue to expand its capacity because wild capture is only expected to increase by 1.5% per year over the next 15 years, failing to meet demand. Furthermore, the public is increasingly demanding that fish be produced using environmentally friendly, sustainable ways. RAS will be the key technology that allows the global aquaculture sector to meet the world's per capita seafood needs over the next decade while being environmentally friendly (Hargreaves, 2006). RAS-produced supply should not be viewed as a competitor to wild-caught fish, but rather as a complement to the oceans' ability to deliver a sustainable level of production that must be complemented by aquaculture-produced goods. This isn't a battle between traditional fishing and the latest technology. To sustain current global per capita consumption levels, an extra 77 billion kg of supplies will be required by 2020 (Ahmad, Sheikh Abdullah, Hasan, Othman, & Ismail, 2021). Due to a known shortfall in fishery-supplied products, rigorous attempts to better manage natural fisheries have been made. The fishery problem is caused by poor management of fishery resources. RAS needs more capital than most other forms of traditional aquaculture systems, and their profitability is based on economic productivity per unit volume of culture space. Support features such as backup generators are included in the RAS farm's overall production expenses. Sufficient production volume must be obtained in order for these support component costs per unit of production to be decreased to a level where overall production costs are competitive. Large-scale production is achieved by repeating a successful modular unit that protects itself against the failure of other production modules. In response to increased market demand, this characteristic allows for controlled, incremental increases in seafood output. It also makes commercial structures like franchising, contract growing, and production-marketing cooperatives easier to raise funds and investment risk. RAS technology allows operators to keep up with changing seafood preferences on the market.

## 2.4. COMPETITION

Long-term competitiveness in the meat market will be largely determined by the cost of feed used to raise the animals and the efficiency with which feed energy is converted into meat (Bess, 2006). The ultimate competitiveness of a meat is influenced by both feed pricing elements and feed conversion efficiency. This is a significant benefit for plant-based fish production, as feed costs will ultimately dominate production costs, while all other cost inputs can be decreased by expanding scale and achieving corresponding production efficiencies. So, for indoor RAS produced fish, feed conversion efficiency combined with productivity per unit area should provide for a considerable long-term benefit in having the lowest feasible production costs. Broiler hens generate roughly 76 kg per square meter of building space per year, while indoor warm-water fish systems with 2.4 m deep tanks create around 290 kg per square meter of building floor area (Scanes, 2018). The success of a commercial aquaculture business hinges on creating the best conditions for rapid expansion at the lowest possible cost of resources and capital. The capacity to regulate the aquatic environment and crucial water quality parameters to improve fish health and growth rates is one of the major advantages of intensive recirculation systems. Despite the fact that the aquatic environment is a complex ecosystem with numerous water quality variables, it is fortunate that just a few of these variables are critical. Temperature, pH, and dissolved oxygen, ammonia, nitrite, $CO_2$, alkalinity, and suspended particle concentrations are all significant characteristics. Each characteristic is significant, but the sum and interplay of all the parameters has an impact on the fish's health and growth rate.

Each water quality metric interacts with and influences other parameters in a variety of ways, some of which are complex. When aeration and degassing problems develop, for example, carbon dioxide levels rise while dissolved oxygen levels fall. Not only is there less oxygen accessible to the fish in this situation, but the fish are also less able to use the oxygen that is available. The fishes' capacity to carry oxygen is harmed by the high carbon dioxide level in the water, exacerbating the stress caused by low dissolved oxygen levels. The link between pH and ammonia toxicity is another example of the complicated interaction among water quality factors. The ionized form of ammonium is far more dangerous than the unionized fraction of total ammonia concentration, and at low pH, most of the ammonia in the water is in the non-toxic ionized form. However, a one-unit increase in pH, from 6.5 to 7.5, raises the hazardous unionized ammonia concentration by a factor of 10 (Tomasso, 1994). Simply increasing the alkalinity of a system with baking soda can result in hazardous quantities of unionized ammonia. The link between water quality indicators and their impact on fish growth and health is complex. Fish lack the ability to regulate their body temperature and maintain it regardless of their surroundings. Changes in temperature alter the rate of biochemical reactions in fish, resulting in varying metabolic and oxygen consumption rates. These rates drop as the species' acceptable temperature range narrows. Fish become active when water temperatures increase, consuming more dissolved oxygen but also creating more carbon dioxide and other excretory products like ammonia. If these parameters are to exceed nominal levels, rising rates of consumption of required elements and generation of harmful components can have a direct impact on overall fish health and survival. Even moderate levels of stress can have long-term negative implications, such as lower growth rates or mortality from opportunistic organisms that prey on stressed fish. Some of the engineering principles that follow require an awareness of certain physical aspects of water. Water has various special qualities that allow life to exist on the planet. At the freezing point, the density drops dramatically, causing a big problem. Ice has a lower density than water and floats. The density of water is also increased by the addition of salt and other contaminants.

## 2.5. VISCOSITY

Water viscosity, also known as absolute or dynamic viscosity, is a measurement of a fluid's r0esistance to shear. When shear force is applied to very viscous fluids like molasses or motor oil, they flow very slowly. At 20°C, absolute viscosity is measured in centipoises, and water has a value of 1.0 centipoise, compared to 0.17 centipoise for air at this temperature (Viadero & Noblet, 2002). As the temperature drops, the viscosity rises. As a result, pumping costs will rise slightly as water temperature drops due to increased viscosity and density. The pressure imposed by a material's gaseous phase when it is in equilibrium with its solid or liquid phase is known as vapor pressure. The pressure at which a liquid begins to boil and turn to a vapor is another way of describing vapor pressure. Pure water's vapor pressure is a function of temperature, and it rises as the temperature rises. The vapor pressure of water is reduced when salt is added to it. Water may change phase from liquid to vapor in a closed pipe due to a decrease in pressure, even if the temperature remains constant. This can happen on the suction side of the pipe, where vapor bubbles form in low-pressure areas and then burst when they reach higher-pressure areas downstream. This is referred to as cavitation.

When it comes to water, having too much is always preferable to having too little. The amount of water required is determined by a variety of factors, including species, density, management practices, production technology, and the level of risk one is willing to take. At a bare minimum, enough water is required to fill production tanks in a reasonable amount of time, provide for routine and emergency tank filter backwashing, facility wash down and clean-up, and home needs. A decent rule of thumb is to have enough water on hand to support a daily water exchange of 100% of the entire system volume. Beyond the minimum exchange volume, the additional amount of fresh water required for any given system is exactly proportional to the amount of used water that is already present in the system. Aside from the obvious reduction in water demand and lower discharge water that must be treated, there are other advantages to reuse supply water as much as possible. Reduced water heating and cooling requirements, for example, are important aspects in the economics of many warm-water species production systems. Furthermore, and this is becoming increasingly essential, recycling of system waters reduces wastewater discharge and the costs associated with the treatment of large discharge flows.

## 2.6. RECIRCULATING SYSTEMS

Fully recirculating systems reuse a higher percentage of total system volume than partial-reuse systems, and partial-reuse systems reuse a higher percentage of total system volume than fully recirculating systems (Pedersen & Wik, 2020). The rate of depletion or accumulation of critical water quality metrics is affected by the amount of water reused. The higher the reuse rate, the more conditioning the reused water requires to return the water quality to target levels. The concentration of dissolved oxygen is usually the most important and first recognized limiting factor that determines the density of fish grown in a system. The concentrations of unionized ammonia and dissolved carbon dioxide levels are the next most critical components. These two variables are inextricably linked. This is owing to the direct impact of dissolved carbon dioxide on pH and the link between pH and ammonia-nitrogen toxicity. As the amount of dissolved $CO_2$ in the system decreases, the pH rises, increasing the toxicity of the total ammonia-nitrogen in the system. The selected reuse system must be able to maintain required levels of dissolved oxygen while also keeping dissolved carbon dioxide, unionized ammonia, and pH below their respective limitation values.

## 2.7. WATER QUALITY

The limiting water quality element in trout and salmonid raceway production systems is usually the dissolved oxygen concentration between racetrack sections, hence serial-reuse systems have been widely adopted (Reid et al., 2009). The water can be reused in the next portion of the raceway by adding oxygen until the accumulated ammonia levels become too high. Due to the requirements for more advanced oxygenation systems and monitoring systems, this basic design has substantially boosted raceway production, but at a greater system cost.

A high-quality water source with sufficient capacity to meet initial needs and anticipated future expansions is one of the most crucial prerequisites for a successful aquaculture plant site. As a result, while selecting a site for aquaculture, the water supply must be properly evaluated and assessed in terms of both quality and quantity.

Aquaculture differs from other agricultural activities in that it necessitates a steady supply of high-quality water. Because most wells are only used on a regular basis, a traditional well pump test is usually only done for a short period of time to determine the well's ability to deliver a specific amount of water in a specific amount of time. The well, on the other hand, is continuously pumped during typical aquaculture operations. As a result, the well's actual output may be substantially lower than the short-term test indicates. This will lead to significant problems down the future (Figure 2.2).

**Figure 2.2.** Water quality.

*Source: https://news.ucsc.edu/2021/07/images/trout_450.jpg.*

Due to the ambiguity about their impact on production, several other water quality criteria are also being thoroughly examined. The impact of high levels of fine solids and organic compounds is unknown at this time, but they have been linked to disease outbreaks.

The leaching of surface-active compounds from uneaten feed and feces, and the resulting depression of surface tension, could be a key design consideration. The corrosion of pipes and fittings, as well as vitamin premix, have the potential to build up dangerous quantities of some heavy metals, such as cadmium, copper, and zinc, in limited exchange systems (Morrow, Reilly, Olivas, & Romero, 2001). The toxicity of heavy metals is highly influenced by the chemistry of the water. These substances can have a significant negative impact on aquatic species by interfering with their endocrine systems, potentially resulting in lowered fertility and development rates.

The ambient chemistry of several EDCs is complicated, and analyzing them is expensive. Dissolved oxygen is the most crucial and vital of all the water quality characteristics, necessitating regular monitoring in intense production systems. Aquatic animals must invest significantly more energy to take dissolved oxygen from water than land-based species must expend to obtain oxygen from the air. Furthermore, when salinity increases, oxygen solubility decreases. The oxygen concentration is also affected by barometric pressure and altitude. Furthermore, these effects are modified by exposure period, fish size and health, water temperature, carbon dioxide content, and other environmental factors. Warm water fish, on the other hand, feed best, grow fastest, and are the healthiest when DO levels are above 5 mg/L (Paperna, 1991).

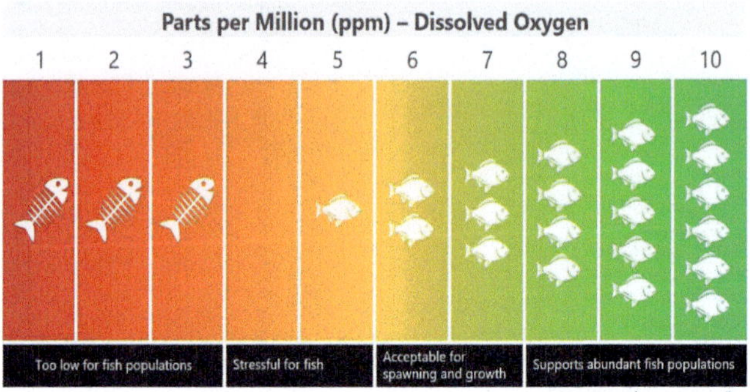


**Figure 2.3.** Dissolved oxygen and aquaculture.

*Source:https://d2gfc6j4v8hvtl.cloudfront.net/wp-content/uploads/2020/02/Dissolved-Oxygen-Healthy-Levels.jpg.*

Dissolved oxygen concentrations (Figure 2.3) above this saturation threshold appear to provide little further advantage to fish. When dissolved oxygen concentrations are at the necessary concentrations, the gills can only transfer a certain amount of oxygen to the blood, and they are extremely close to or at their maximum transfer capabilities. Higher oxygen levels in the water have no effect on the amount of oxygen carried by the bloodstream.

Water temperature is only second to dissolved oxygen in terms of importance and impact on a commercial aquaculture's economic sustainability. Temperature has a direct impact on physiological functions such respiratory rate, eating, and assimilation efficiency, development, behavior, and reproduction (Barton & Iwama, 1991). Cool-water, and warm-water fish have traditionally been divided into three categories based on their preferred temperature. Fish are classed as poikilothermic, or cold-blooded, meaning that their body temperature is similar to that of their surroundings. As a result, each species has an optimal temperature range for growth, as well as an upper and lower limit beyond which it would die. Growth rates increase as the water temperature rises within the species' tolerated temperature range, until the optimal temperature is attained. Furthermore, at temperatures above the optimum, the fish food conversion rates are reduced. To ensure maximal growth while minimizing stress, system temperatures must be kept as close to optimum as feasible. Nitrogen is found in proteins, nucleic acids, adenosine phosphates, pyridine nucleotides, and pigments, and is an important nutrient for all living organisms. Nitrogen is only required in less amounts, and physiological requirements are easily met. Excess quantities generate nitrogenous wastes, which must be removed. Through gill diffusion, gill cation exchange, and urine and feces excretion, the fish produce and expel diverse nitrogenous waste products.

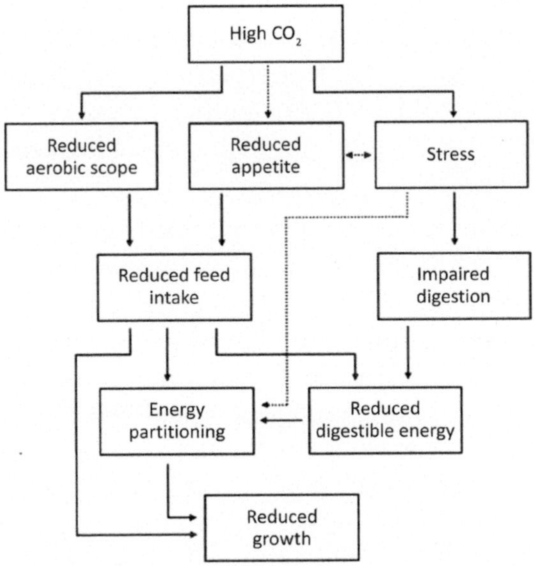


**Figure 2.4.** Carbon dioxide and aquaculture.

*Source:https://ars.els-cdn.com/content/image/1-s2.0-S1546509819300044-f08–04–9780128176092.jpg.*

## 2.8. CARBON DIOXIDE

Carbon dioxide is extremely soluble in water, but because of its low concentration in the atmosphere, concentrations in pure water are low (Figure 2.4). Animal respiration and organic matter breakdown produce the majority of the carbon dioxide in an aquaculture water column. Depending on the biological activity in the aquifer recharge zone, the concentration of carbon dioxide in groundwater can range from 0 to 100 mg/L (Sanni & Forsberg, 1996). The rate of respiration, photosynthesis, and gas exchange with the atmosphere influence the concentration of dissolved carbon dioxide in surface waters. High carbon dioxide concentrations impair respiration efficiency and diminish tolerance to low dissolved oxygen concentrations. Carbon dioxide excretion at fish gills is reduced when there is a lot of it in the water. As a result, the $CO_2$ content in the fish blood rises, reducing the blood plasma pH and resulting in a condition known as respiratory acidosis. Even with high concentrations of DO in the water, the amount of oxygen that the blood hemoglobin can carry is reduced in this environment, and respiratory discomfort might result. The Bohr-Root effect is the name for this phenomenon. Higher concentrations have a narcotic effect on aquatic animals and are occasionally used as a temporary anesthetic to alleviate handling stress and during treatment procedures as a management strategy (Rummer & Brauner, 2011). High quantities of carbon dioxide have also been used to induce fish to swim through connecting pipes from one tank to another.

Fresh, brackish, and salty are all terms used to describe water. Each of these categories refers to the salinity of the water, and there are no clear distinctions between these classifications. The overall concentration of dissolved ions in water is known as salinity, and it is measured in parts per thousand (ppt) or grams of salt per kg of water. Natural water salinity is influenced by the temperature, topography, and hydrological conditions of the immediate environment. Although the salt tolerance of most aquaculture species is rather extensive, each aquatic species has an optimum range of salinity for reproduction and growth. Rainbow trout fingerlings, for example, are raised in freshwater, acclimated to saltwater, and then developed to marketable size in marine cages with salinities as high as 20 ppt. Most commercially important freshwater fish breed and grow successfully at salinities of at least 4–5 ppt (Ghittino et al., 2003). Fish regulate the intake of ions from the environment and limit ion loss to maintain the concentration of dissolved salts in their bodily fluids. When aquatic organisms are subjected to salinity levels outside of their optimal range, they must devote a significant amount of energy to osmoregulation at the sacrifice of other processes, such as development. If the salinity falls too much below the ideal, the animal loses its equilibrium and dies.

The capacity of an aqueous system to emit or take electrons from chemical reactions is measured by the Oxidation Reduction Potential or ORP. ORP, like pH, is used to characterize a system's water quality, although it's most commonly utilized in aquaculture to control ozone dosing. The ORP sensor uses a two-electrode setup to make a potentiometric reading expressed as either a positive or negative millivolt, similar to a pH probe (mV). A solution with a positive ORP measurement gains electrons, while a solution with a negative ORP measurement loses electrons. Despite the fact that ORP measurement is very simple, discrepancies in laboratory and in-plant sampling can be significant due to the effects of solution temperature and pH, slow electrode kinetics, electrode poisoning, and tiny exchange currents, which cause drift and poor responsiveness. Still, rather than assessing their absolute value, ORP measurements have proven beneficial as an analytical tool for monitoring changes in a system, such as managing ozone dosing.

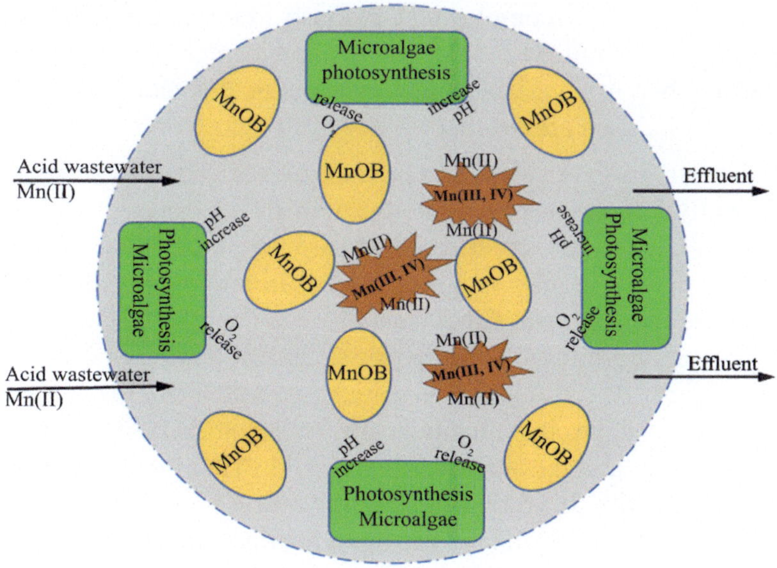


**Figure 2.5.** Mn and aquaculture.

*Source:https://www.mdpi.com/water/water-11–02493/article_deploy/html/images/water-11–02493-g001.png.*

## 2.9. MANGANESE

The manganous ion (Figure 2.5) is oxidized to manganous dioxide by dissolved oxygen in these alkaline circumstances (Axler et al., 1996). The precipitate is subsequently dissolved in sulfuric acid, which creates acid conditions for the oxidation of iodide to iodine by manganous dioxide. This may appear complicated, but in practice, all of the reagents are pre-mixed in many widely accessible test kits, making DO measurement a common occurrence. Due to the low cost of dissolved oxygen meters, this method is typically exclusively utilized for cross-checking and calibration purposes. The technique by which oxygen is carried into a fish culture vessel and waste materials created within are removed is known as water flow. The main elements impacting water quality and fish production, such as oxygen, ammonia, carbon dioxide, and suspended particles, should be carefully balanced in the design of a RAS. To determine the thresholds for each of these parameters, the value of each must be calculated separately. The system must then be run at the lowest flow rate possible while maintaining a particular parameter at its design value, such as ammonia, after the relevant calculations have been completed. Obviously, the lowest flow rate that may be achieved while preserving one parameter may be insufficient for sustaining another. Any variable affecting water quality can be studied using the same mass balance approach.

Smaller fish have a tendency to be overstocked. Tilapia can be stocked in significantly larger numbers while the animals are young. Crowding, on the other hand, is dangerous since it can harm small animals in their rapid growth stages, such as fish less than 13 cm in length. Because you will only see the active dominant fish, visual observations can be deceiving. The best way to tell if your stocking densities are too high is to sample fish populations. Trout raised at higher densities are more likely to have their fins degraded. Wide changes in population size, on the other hand, may be due to poor water quality and less-than-optimal feed distribution efficiency rather than stocking density. If you use a sinking feed, for example, you may be able to stock more heavily than if you use a floating feed. Each time the fish is relocated, they will lose weight and die as a result of the trauma. Although the time and labor required to accomplish the movement are obviously essential concerns, it is this loss of productivity that limits the number of migrations from an economic aspect. Other management approaches to enhance the RAS system's utilization include connecting numerous tanks to a common water treatment system and keeping the fish in the same tank they were stocked in. Any fish movement is undesirable and should be avoided by design wherever possible. If it's not possible to prevent it, make every attempt to keep the transition as stress-free as possible. Flumes or raceways connecting individual tanks, as well as proper design and architecture of the grow out facility, can help achieve this. When compared to raising the same quantity of fish in smaller tanks, producing food fish in big circular tanks has resulted in significant cost savings. Larger circular culture tanks have numerous advantages in the production of food fish. Indeed, the capital expenses of large vs. small tanks are not proportionate to tank volume. Using larger tanks can result in significant cost reductions (Michael B. Timmons, Summerfelt, & Vinci, 1998).

Because the overall rate of water exchange in small tanks is rapid, large tanks are more critically dependent on tank hydraulic design than small tanks. Because the high turnover rate brings more oxygen into the tank and flushes wastes quickly, the rapid hydraulic exchange can give good water quality. Throughout big tanks, on the other hand, the HRT is typically longer and as a result, the inlet and outlet injection methods, as well as the flow rate, become the most important elements influencing the uniformity of water conditions in the tank. Water exchange, feeding rate and waste output all affect a tank's carrying capacity. In contrast to raceways, water velocities in round tanks can be controlled to a great extent regardless of the water exchange rate. The design of the inlet and outlet, as well as the flow through the tank and the central drain, are the most important aspects. Self-cleaning properties are further aided by properly constructed intake-outlet arrangements. Compared to raceways, circular tanks provide a number of significant advantages. Round tanks are less expensive, and they may be used at low water exchange rates while still producing ideal velocities, hydraulic patterns, and self-cleaning properties. It is often possible to achieve the greatest cumulative oxygen consumption (COC) in a single culturing unit using circular tanks (Benetti et al., 2001). In other words, instead of flowing through two or three tanks, the water is used after passing through just one.

**Figure 2.6.** Circular tanks.

*Source: https://5.imimg.com/data5/UA/GD/KM/SELLER-4414671/round-shape-aquaculture-farm-tank-500x500.jpg.*

## 2.10. CIRCULAR TANKS

Circular tanks (Figure 2.6) can achieve nearly complete mixing, which means that the concentration of dissolved ingredients in the water changes instantly to the concentration in the tank. As a result, if proper mixing is done, every fish in the tank will be exposed to the same water quality. By optimizing the design of the water input structure and selecting a water exchange rate such that the limiting water quality parameter does not reduce production when the system achieves carrying capacity, good water quality may be maintained throughout the circular culture tank (Behroozi & Couturier, 2019). The water intake and outlet structures, as well as fish grading and removal mechanisms, should be designed to reduce fish handling labor and achieve uniform water quality, rotational velocities, and solids removal inside the circular tank. Water flow is injected tangentially to the tank wall at the tank outer radius, causing the water to spin around the tank center, providing a primary rotating flow. The no-slip condition between the primary flow and the tank's bottom and side walls produces a secondary flow with an inward radial flow component at the tank bottom and an outward radial flow component at the tank surface. This inward radial flow along the tank's bottom takes settleable solids to the center drain, giving circular tanks their intended self-cleaning ability. Unfortunately, in a circular tank with this flow, one portion around the center drain region will become a no-rotational zone with lower velocities and poor mixing, causing sediments to settle on the floor. The diameter to depth ratio, the introduction of tangential flows near the walls, and the overall rate of flow leaving the center bottom drain all influence the size of the no-rotational zone.

**Figure 2.7.** Raceways.

*Source: https://ars.els-cdn.com/content/image/3-s2.0-B9780124095489037647-f03764–02–9780124095489.jpg.*

Raceways (Figure 2.7) are most commonly connected with salmonid culture. Attempts to adapt raceway culture methods to RAS have been made in the past, with varied results. Raceway supporters would typically argue that raceway design makes greater use of floor space and allows for easier handling and sorting of fish (Huggins, Piedrahita, & Rumsey, 2004). Their downside is their inability to self-clean quickly under normal working conditions, which increases nutrient and BOD dissolution into the flow. This clearly illustrates that any solid surface screening material, in particular, will require regular cleaning. If at all possible, avoid putting items in fish tanks. As a result, the finest water quality is found at the top of the tank, where the water enters, before gradually deteriorating along the raceway's axis towards the outflow. It's not uncommon to witness fish crowding towards the raceway's intake end in order to get the most oxygen, or near the raceway's exit end to avoid high concentrations of some other accumulating pollutants. Because it is more difficult to disperse feed throughout raceways than it is in circular tanks, fish will cluster around the feeders during feeding activities. The average velocity through the raceway is 2–4 cm/s, which is insufficient to efficiently remove settled solids from the culturing space (Stucchi, Sutherland, Levings, & Higgs, 2005).

Water is pushed upwards from an indoor fish tank, through the distribution plate, and then falls through the rectangular chambers, flowing over the dam boards at the end of a raceway. The gas-liquid interface required for mixing and gas transmission is provided by these chambers. The streams of falling water impact a gathering pool at the bottom of each chamber, where the effluent water flows away in equal amounts from each chamber. Pure oxygen is injected into the first rectangular chamber, then passed through a sequence of separate chambers before being vented to the atmosphere at a significantly lower concentration.

Each of the rectangular chambers is sealed, and the single holes between them are placed correctly to prevent back-mixing. Raceways provide various advantages over circular tanks, including easier sorting, grading, and handling of fish, as well as better floor area use. At various different exchange rates, the results of these investigations revealed good bottom velocities for scrubbing particles and pushing them towards the center drains in each cell of the prototype mixed-cell raceway. Mixed-cell raceways have a lot of promise as a retrofit for existing raceways or as a new production system design. In culture tanks, production efficiency can be improved by raising the carrying capacity (Labatut, Ebeling, Bhaskaran, & Timmons, 2007).

As the fish consume dissolved oxygen during respiration, dissolved oxygen is frequently the first water quality measure to limit culture tank carrying capacity. In fish culture tanks, the amount of dissolved oxygen available is determined by the water flow rate multiplied by the available dissolved oxygen concentration. If water is not limiting and the dissolved oxygen concentration remains constant, doubling the water flow, or hydraulic exchange rate, through a culture tank will double the tank's carrying capacity. Moving more water through the culture volume, on the other hand, is costly since it necessitates large pumps, pipes, or high-water pressure, which may not be possible, and will raise the farm's fixed and variable expenses. Supersaturating the DO in the water flowing into the tanks, on the other hand, has become popular and is typically a more cost-effective way to boost the profitability of tank-based fish farms. Year-round fish stocking and harvesting can achieve the highest economic productivity of the culture system because continuous production keeps the culture system at or just below its carrying capacity. Furthermore, harvesting at a specific size boosts product value by supplying the market with more consistently sized fish.

Continuous stocking and harvesting tactics necessitate rearing multiple size groups at once, frequent handling of the fish, and better inventory accounting techniques. If automated equipment is not employed, this might stress the fish and increase labor costs. A practical system for size sorting the fish, counting them, and moving them to other areas should be built into the culture tank and facility design to minimize the cost of handling and grading fish to a minimum, as well as the stress on the fish to a minimum. Many hatcheries use hand grading devices like box graders, which have proven to be effective on small farms where the expense of automatic grading and counting equipment is prohibitive. Mechanically driven belt graders and roller graders are common automatic graders. As the fish travel through the size mechanism on these mechanical graders, they are normally taken out of the water for a short time. Despite the fact that mechanical graders cause some stress and damage, many of the well-established commercial mechanical grading equipment are regarded as safe, dependable, and quick ways of size sorting and counting vast numbers of fish. With all new technology, it is best to check with other fish farmers who have tried the equipment before acquiring it. Other fascinating and novel technologies that use ultrasonic, infrared light, and video systems to assess fish size distributions within culture tanks and cages are already commercially accessible. When feed input and fish numbers are known, this type of inventory tracking technology can be used to measure fish growth, feed conversion, and predict the fraction of fish reaching harvest size. If a circular system is required owing to space or other constraints, the system's design must recognize and accept the same functions that can be achieved more simply in rectangular systems. An inlet zone, a settlement zone, and an outlet zone are all required. However, unlike rectangular systems, the boundaries between these zones are not easily discernible. The arrangement of inlets, the hydraulic movement of the water inside the tank, the physical qualities of the solids within the water, and the placement of the drain and outlet pipes all contribute to the formation of these zones within the circular tank.

The flow towards the outlet is regulated in these conditions to cause minimal turbulence, especially near the outflow pipe in the tank. A circular barrier that surrounds the outflow pipe can be constructed to keep floating solids out of the settling tank. To boost the level of treatment, granular-medium filters are frequently employed in conjunction with other treatment techniques. Solids, phosphorus, algae, turbidity, and pathogens are all removed more effectively with these filters. The water is pre-treated for phosphorus removal by adding chemicals that precipitate phosphorus, such as alum or ferric chloride. The granular-medium filter then removes the precipitated phosphorus in its particulate form. It's worth noting that the backwash from these filters will include a lot of total phosphorus. All of these filter types operate on the same principles; they all use the previously mentioned removal methods and require a backwash or cleaning process to regenerate the filter's particulate removal capacity.

The continuous backwash deep-bed up flow filter, for example, equally introduces water at the filter bottom, where it flows through a sand bed. The filtered water exits the sand bed, passes through an overflow barrier, and exits the filter. Sand and solids trapped by the sand are dragged downward within the sand bed and into the suction of an airlift chamber in the center of the filter during this filtration. The sand and solids are carried to the top of the filter and deposited in a reject compartment by the airlift. Because the reject compartment's discharge barrier is lower than the filtrate dam, the reject compartment receives a minimal flow of filtered water that transports particles out of the filter on a continuous basis. This backwash flow also cleans sand that has been flown to the reject chamber and is now settling back down to the top of the sand bed.

**Figure 2.8.** Foam.

*Source: http://vietlinh.us/nuoi-trong-thuy-san/ikt/ao-tom-vang-bot.jpg.*

## 2.11. FOAM

Foam removal is usually an unexpected, and unreliable operation (Figure 2.8). However, if you are persuaded that you must remove the foam. It's important to remember that foam fractionation isn't a substitute for good primary TSS removal. The presence of surfactants, which are proteins, causes foaming (M. B. Timmons, Chen, & Weeks, 1995). As a result, a dead fish crushed with a pump or left in the tank to degrade will give a considerable amount of surfactant, resulting in immediate and apparent foaming. Environmentally appropriate waste management and disposal will become increasingly crucial in all agriculture/aquaculture operations, particularly marine aquaculture, as environmental restrictions become more stringent. The handling of solid wastes generated by aquaculture systems involves particular storage and disposal challenges due to their high moisture content. Suspended particles and phosphorus in the released wastewater are two major problems. Mechanical filters such as microscreens, sand filters, and floating bead filters provide a separate solid waste effluent stream discharge that benefits from further concentration to lower the quantity and enhance the quality of the discharge. To save salts and decrease or eliminate the environmental impact of salt discharges, it is necessary that a considerable portion of the waste effluent stream be recycled back into the production system for marine RAS. Furthermore, thickening wastes through dewatering expands the range of final disposal substitutes while reducing the volume and expense of storage and transportation. Improved waste management systems will increase the economic viability and sustainability of aquaculture systems, hence developing better systems for disposal and utilization of aquaculture wastes is a top priority. Aquaculture wastes should be used as an environmentally friendly product whenever possible.

Land application and composting are the two most prevalent ways for recycling solid waste from aquaculture facilities. Most states have standards or regulations that limit the quantity of pathogens, heavy metals, and other pollutants that can be applied to agricultural crops when manure and other organic wastes are used to fertilize them. To avoid runoff or groundwater contamination, land application rates are depending on nutrient content, soil type, and plant nutrient uptake characteristics. In developed locations, odor issues can also limit land use. Finally, because aquaculture sludge is primarily water, transportation from the point of production to the application location can be a significant element in sludge management expenses. Sludge and effluent wastewater must first be moved to a storage system for sludge thickening and flow equalization before either of these disposal options may be used (Jasmin, Syukri, Kamarudin, & Karim, 2020). The hydraulic loading on future operations will be reduced as a result of sludge thickening. During cleaning and other maintenance actions, sludge flow rates and concentrations will also change. Anaerobic lagoons, earthen ponds, and above and below ground tanks are examples of sludge storage facilities. Sedimentation basins, geotextile bags, belt filters, lime stabilization, wetlands, and sand beds can all help thicken sludge. To facilitate settling and, in some situations, sequester dissolved phosphorus, coagulation, and flocculation aids are frequently used. Captured solids from solids removal procedures are typically dilute, containing less than 2% total solids. In settling basins, these solids can be concentrated or thickened to a total solids content of 5–10%. By lowering the volume of sludge, solids thickening also lowers disposal costs.

CHAPTER 3

# Aquaculture in Urban Areas

CONTENTS

## 3.1. INTRODUCTION

The world's urban population will reach 5.1 billion people by 2025, equaling the total human population on the earth in 1930. At regional and global dimensions, this vast trend of landscape urbanization will have significant impacts on ecosystems (Figure 3.1). The worldwide population will expand by 50%, global economic activity will increase by 500%, and global energy consumption and manufacturing activity will increase by 300% (Trottet, George, Drillet, & Lauro, 2022). If we delay the essential planning and engineering needed to assure the sustainability of water resources for both nature and millions of people, we risk significant biodiversity losses and the entire destruction of the remaining intact coastal ecosystems. Given these obstacles, some researchers advocate for a reduction in the rate of human impacts, as this drives all resource use and waste generation; a reduction in the rate of human impacts, as ecosystems can respond to lower rates of impacts and stabilize with moderate levels of human-induced changes; and an acceleration of ecosystem-level research and management.

**Figure 3.1.** Urban aquaculture.
Source: https://d178ivhysawugh.cloudfront.net/media/l/920x561/1519478386/singapore.jpg.

Coastal waters are among the most frequently used and modified environments on the planet, with ecosystem destruction and pollution at a high rate. Sewage, solid wastes, refuse, sediments, dust, pesticides, and oil hydrocarbons are all delivered to rivers, estuaries, and oceans by human activities in the coastal zone. Around 80% of marine pollution comes from land-based sources and activities, according to estimates (Chua Thia Eng,

Paw, & Guarin, 1989). A massive global effort is underway to determine the extent, quantity, and delivery of pollution from land-based sources to coastal waters. Humans exploit more of the world's freshwater runoff, making them one of the key driving forces of the Earth's hydrologic cycle (Figure 3.2).

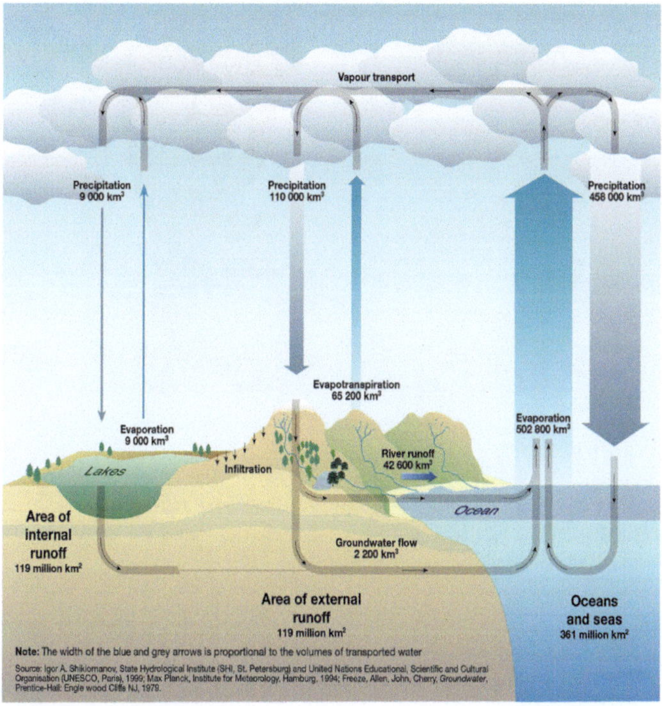


**Figure 3.2.** Hydrologic cycle.

*Source: https://farm6.staticflickr.com/5574/32323667746_de3aebca43_o.jpg.*

Water is becoming increasingly scarce in areas with low rainfall, such as the Middle East, Africa, and parts of Asia. Many major rivers are overused to the point where little or no water reaches their deltas and the sea. Only 2% of rivers in the United States are free-flowing, while several important inland water regions such as the Aral Sea and Lake Chad have nearly dried up or been reduced to small, saline lakes (Siddiqui, Conkle, Scarpa, & Sadovski, 2020). By 2025, it is likely that two out of every three people will live in water-stressed areas, with 450 million people in 29 countries now experiencing water scarcity. By 2025, it is expected that 25 African countries will be experiencing water scarcity. When exploring for new sources of water in water-scarce areas, wastewater recycling is one of the most popular solutions (Fletcher, 2022).

**Figure 3.3.** Water scarcity.

*Source: http://www.fao.org/fileadmin/user_upload/zhc/img/24635_0135_620x338.jpg.*

## 3.2. WATER

While water scarcity is a key concern in water-scarce areas, water quality concerns are a global concern (Figure 3.3). In many regions of the world, clean water supplies and sanitation are serious issues, with 20% of the worldwide population lacking access to safe drinking water. In underdeveloped nations, water-borne infections caused by fecal pollution of surface waters are a major cause of disease. Polluted water is predicted to have a negative impact on the health of 1.2 billion people and is responsible for the deaths of 15 million children each year (Cao et al., 2007). Nutrient inputs from wastewater, development, and agriculture to the coastal zone represent a serious danger to the long-term viability of coastal marine ecosystems. Increased wastewater discharges cause severe environmental changes in nearshore areas, have a negative impact on human health, and have significant economic implications for tourism, recreation, fishing, and aquaculture. Wastewaters, on the other hand, are significant sources of nutrient-rich freshwaters, resources much too valuable to be squandered by simple disposal and dilution in the world's oceans. To reduce water consumption, increase the quality of receiving waters, and reduce overall environmental impacts, treated wastewater should be reused for agricultural and industrial applications as part of a holistic urban planning process.

**Figure 3.4.** Eutrophication.

*Source: https://www.conserve-energy-future.com/wp-content/uploads/2017/01/eutrophication-hypoxia.jpg.*

## 3.3. ECOSYSTEMS

In marine ecosystems, nitrogen is the key limiting factor for plankton production. Eutrophication (Figure 3.4) symptoms are caused by significant increases in nitrogen and sediment in coastal waters. Anoxic conditions causing catastrophic impacts on fish and invertebrates increased abundances of other species such as filamentous algae, macrophytes, and jellyfish. There are numerous examples of increased wastewater discharges having negative effects on coastal ecosystems. Increased nitrogen loadings have been linked to algal blooms and the degradation of the structure and function of coastal marine ecosystems off San Pedro Bay. Similar findings have been reported for the Baltic Sea coast of Sweden. Natural turbidity inhibits light penetration in some estuaries, which limits phytoplankton responsiveness to nutrient increases, whereas phytoplankton blooms themselves limit light penetration in others. Nutrients are transported from the Hudson River and New York Harbor where light penetration is limited and nutrient input is high causing impacts to offshore ecosystems rather than estuaries and adjacent nearshore ecosystems that are directly impacted by wastewater discharges.

Light constraint caused by phytoplankton blooms (Figure 3.5) in response to fertilizer inputs has been linked to the decline of eelgrass, an important bay macrophyte in Chesapeake Bay. Nutrient inputs, when used in moderation, can boost primary production in coastal waters, resulting in more food for shellfish and other filter-feeding species, and hence more productive marine food webs. Primary productivity stimulation can be used to supply food for juvenile fish, which can then be cultured to produce food or fed to target culture species.

**Figure 3.5.** Phytoplankton blooms.

*Source: https://www.globalseafood.org/wp-content/uploads/2015/12/Green_phyto_in_shrimp_pond1–337x256.jpg.*

As the world's population grows, so will the need for drinkable water, and because water is a restricted, finite resource, planning for water recycling and reuse will have to become mainstream as a cultural standard. In some nations, wastewater is already utilized to irrigate crops: in Israel, 60% of total effluent is utilized, 25% in India, and 24% in South Africa. It has been reported that using wastewater to irrigate crops can enhance yields by up to 30% (Adeleke, Robertson-Andersson, Moodley, & Taylor, 2021). Golf courses in France are becoming important users of wastewater for grounds-keeping. The WHO has framed a set of standards for the reuse of wastewaters. While water markets have evolved, they have not progressed quickly enough to prevent the rising discharge of wastewaters into the ocean. Even though reclaimed water in Los Angeles County is provided at a significant discount compared to potable water, this situation arises. Wastewater storage reservoirs are a viable alternative in arid and semi-arid locations when access to natural water supplies limits agricultural productivity. Wastewater is pumped into storage reservoirs in Mexico before being used for agriculture. Raw wastewater effluent is fed into a storage reservoir and held there for up to seven months. The effluent is then transferred to a second reservoir, where it is held for another 2–6 months, achieving a higher grade that has been deemed safe for irrigation by WHO, as long as the palatable portion of the crop is not in direct contact with irrigation water. Wastewater from the second reservoir was found to cause diarrhea, especially in children, when it was sprayed to crops that are directly ingested.

**Figure 3.6.** Artificial lakes.

*Source: https://image.made-in-china.com/202f0j00QMfWlskcbZoY/HDPE-Geomembrane-Film-for-Agriculture-Aquaculture-Use-Geomembrane-for-Artificial-Lake.jpg.*

## 3.4. ARTIFICIAL LAKES

Artificial lakes (Figure 3.6) were built in the Netherlands to capture river water, which was subsequently pumped to underground aquifers that had become overdrawn due to urbanization. While the Netherlands' example does not include wastewater treatment, comparable techniques may be used to recycle treated wastewaters into aquifers for reuse in urban habitats (AbuDalo et al., 2021). Agroindustry refers to a diverse range of businesses that process or fundamentally change agricultural raw resources. Despite the fact that defining urban aquaculture is nearly as complex as defining aquaculture itself, many varieties of urban aquaculture fall within the agroindustry. Raw materials like feed and fingerlings are turned into value-added products, whether they're living organisms or fresh filets. The complexity of the transformation processes used by traditional agroindustry varies, and the more complicated the process, the larger the capital, technical, and management required. The same patterns can be seen in urban aquaculture operations. Much of this effort has been done by international organizations like the World Bank. Identification of opportunities, project design, project analysis, and implementation are all steps in the agro-industrial development process. Every year, universities around the world generate a large number of experts with either industrial or agricultural competence, but just a handful with both. The similar focal separation can be seen in governmental entities and multinational organizations. These disciplines are rarely brought together to the level of integration required for competent agro-industrial project analysis.

## 3.5. AGROINDUSTRY

Agroindustry (Figure 3.7) come in a wide range of types. Cleaning and packaging fresh vegetables, potato storage and packing factories, sugar mills, and chicken production and processing are just a few examples. Despite their variations, they all have a lot of qualities in common. Seasonality, perishability, and quality fluctuation are all present in the raw materials they change to various degrees. The finished goods they sell, on the other hand, have a more consistent demand. This generates a fundamental distinction between agroindustry and typical manufacturing, as well as more issues in inventory management, supply, and demand balance, production scheduling, and internal acquisition, process, and marketing coordination. In the case of urban aquaculture, agroindustry characteristics are typically evident, particularly in terms of seasonality of inputs. Fry and fingerlings are frequently only available for a limited time each year. Production procedures, storage facilities, and scheduling must all be tailored to the needs of the business. In North America, one example is the development of hybrid striped bass in recirculating indoor systems. Fingerlings from this artificial cross are only available for two months each year, hence several large producers have created the notion of cold-banking, which entails bulk housing of fingerlings at low temperatures until production facilities and scheduling can allow their growth. Despite the fact that this approach necessitates considerable investments in infrastructure, labor, and energy, it is required to supply the constant demand demanded by target markets (Martínez-Córdova, Martínez-Porchas, Emerenciano, Miranda-Baeza, & Gollas-Galván, 2017).

**Figure 3.7.** Agroindustry.

*Source: https://www.aquaculturealliance.org/wp-content/uploads/2018/02/Aquaculture_foursquares_Rabobank-960x640.jpg.*

Unfortunately, the possibility of losing money due to disease or technological failure is always present. Perishability difficulties, which are common in agroindustry, are also common in urban aquaculture. Critical inputs like feed and fingerlings are frequently perishable, and storage space is generally limited. Inputs are frequently quite varied as well. Feed quality and palatability, fingerling size and health status, chlorine-chloramine concentration of municipal water supplies, and so on are only a few examples. Indeed, energy supply must be considered variable in many intensive aquaculture facilities, necessitating on-site generators and switching devices. Water will most likely become a key raw material expense in urban aquaculture procedures in the future. One significant difference between urban aquaculture and the general agroindustry model is that, while raw materials are typically the most important cost components in agroindustries, cost factors such as labor, energy, liquid oxygen etc., may partially or completely conceal them in an urban aquaculture enterprise. Planning and management in agroindustrial production can be exceedingly difficult when compared to traditional culturing. Less standardized inputs place more strain on production schedule, cash flow, and quality control. As a result of these issues, the company requires a considerably higher level of integration and information flow. These differences are emphasized in an urban aquaculture operation as different batches of fish go through the production cycle, necessitating segregation, grading, and upkeep until they are harvestable. To avoid putting unnecessary strain on production stocks, space, feed, and labor must all be distributed effectively. Agroindustries are particularly unusual in that they rely on a variety of different economic sectors, ranging from agricultural inputs to service industries and distribution businesses. Each of these connections has the potential to cause disruptions in day-to-day operations, particularly in an urban aquaculture operation.

Despite the inherent risks of these operations, few other investments can compare to well-planned agroindustries in terms of generating economic value. This assertion, in most situations, also applies to urban aquaculture. The combination of aspects from both agricultural and manufacturing analyzes is required for effective planning and evaluation of agroindustrial projects. Nonetheless, fundamental parts of manufacturing and agricultural analysis have been effectively linked on a number of occasions to enable for the design and implementation of well-thought-out agroindustrial firms. Seasonality can be very relevant in the procurement of fingerlings, even though it is less of a consideration for feed inputs. As previously stated, many aquatic species only have fry or fingerlings accessible for a limited time each year. Maintaining seed stock for year-round growth and harvests necessitates specialized storage and grading facilities, which increases the danger of catastrophic losses. When fry and fingerlings must be held for months before being grown out, the health state of the fry and fingerlings upon delivery becomes even more critical. For many intensive aquaculture plants, finding suppliers prepared to take on some of the management and risk associated with extending the period of availability for adequately grown fingerlings might be a top priority.

Input costs and cost determinants are another essential part of procurement analysis. Supply and demand relationships, opportunity costs for operations producing fingerlings, procurement system structure, and logistics will all have an impact on input costs. For operational and planning purposes, spot pricing may be employed, in which a going market price based on numerous suppliers is used to establish an input cost at any given time. Contract pricing from designated suppliers is an option, and it can be based on spot prices, a cost of production plus a particular premium, or a fixed base price with an additional charge based on certain economic and market considerations. In urban aquaculture operations, such arrangements have been successful for both fingerlings and feed supplies. With enough lead time and commitment, favorable costs can often be obtained. Contract pricing clearly enables for the establishment of more long-term procurement agreements, but it also increases reliance on external entities. The goal of procurement cost analysis is to describe and accommodate historical volatility in input costs while also seeking to identify the factors that have led to that variability.

## 3.6. AQUACULTURE

In aquaculture grow out operations, the organization of the fry-fingerling procurement system demands specific attention. It is necessary to assess the number of producers, their capacities, and their transportation capabilities, as well as their geographic positions. Because of competitive pressures or developing opportunities, individual operations may cease to manufacture the species in question, the product mix of each provider becomes an important factor. The ownership patterns of significant suppliers are critical indicators of their stability and should not be neglected. The flow of fry or fingerlings from producers to grow out facilities, whether directly or via live-haulers and brokers, must be examined in order to identify procurement power centers. Both the volume of total production handled and the profits realized between suppliers and customers are indicators of power among particular entities at each step of the procurement pathway. From a logistical and financial standpoint, it may make more sense to adopt backward vertical integration and develop in-house facilities for fingerling production in some cases, particularly for aquatic species that can be reared entirely in recirculating systems.

## 3.7. PLANNING

Analysis and planning activities might focus on the production processes that will be used after a clear understanding of procurement systems for essential inputs emerges. The technology to be used, facility site requirements, inventory management methods, required supplies and inputs, process control and programming, and possibly valuable by-products will all be important factors to consider. When planning and evaluating alternative production technologies, a number of factors must be considered, including labor vs. capital requirements, energy requirements, raw material requirements, and productive use of available capacity (Gentry et al., 2017). The production process must be evaluated from both a financial and economic standpoint. While high levels of automation and process control may maximize return on investment, efforts to shift technology more toward the utilization of manual labor may make more sense if the main goal is local or regional economic growth. In most cases, these two requirements must be balanced to some extent in order to obtain both investment and political support on the local level. Access to supplies, markets, and labor are typically disregarded or undervalued when it comes to facility placement requirements in urban aquaculture growth. Utilities and water supply are also frequently forgotten. Site expenses for urban aquaculture are frequently overvalued, especially when they should be seen as an annualized cost component for any agroindustrial enterprise. While abandoned manufacturing or warehousing facilities may appear to be well-suited for conversion to urban aquaculture at first instance, they are fundamentally problematic constructions in terms of drainage, tolerance of high humidity and salts, and biosecurity.

When determining technology and site needs, the availability of working cash and sufficient lines of credit should be crucial factors. While systems analysis is based on transformation, procurement, and marketing activities, financial, and economic analyzes are also essential for sound agroindustry planning and evaluation. After the procurement, processing, and marketing elements have been analyzed, more standard budgeting methods such as whole farm budgeting. Financial evaluations are normally required for access to financing, whereas economic analyzes are more relevant to local governments and communities where prospective agroindustries will be located. Financial analyzes are mostly concerned with things like return on investment, whereas economic analyzes are concerned with the costs and benefits to local, regional, and national economies. Each has its own format and important factors to consider. Financial and economic studies, on the other hand, cannot be developed successfully until the systems analysis is completed, and both must often show favorable results in order for the implementation phase to begin. Rainbow bass is in high demand, with restaurants and fancy seafood markets competing for it. The retail side of the business has some demand flexibility, but restaurants require regular supplies and consistent quality. Several well-known rainbow bass farmers in the city in question sell their product through huge seafood brokers. Because these growers don't harvest and distribute on a regular basis, long transportation periods and associated stress might cause quality difficulties for both retailers and customers. The planned facility's market advantages would include the ability to provide on a regular basis, respond quickly to unique client demands, and designate the product as locally grown.

Every year in the autumn, rainbow bass have a limited spawning season. Fingerling supplies vary depending on weather and other variables. In some years, they are abundant, while in others, they are scarce. Hatcheries prefer to sell advanced fingerlings trained on feed to grow out operations directly. When nursery facilities are full, hatcheries will often sell a part of their fry to four or five enterprises that will buy them as soon as they become available and raise them to fingerlings for sale. One of these operations has huge, diverse facilities for a wide range of species and often controls around half of the fry-to-fingerling supply chain. Clearly, the hatchery facilities must choose between the extra cash that can be generated by selling fry that would otherwise go to waste and the increased availability of fingerlings on the market and the resulting price drop. In either scenario, fry is only available for 3–4 weeks per year, and fingerlings are only available for up to 8 weeks. Prior to the spawning season, hatcheries will accept deposits from people who want to reserve orders at a predetermined price on a first-come, first-served basis. Fry-to-fingerling operations are less likely to strike such deals, at least in terms of pricing. Several hatchery operators claim fixed capacities as a limit on the availability of advanced fingerlings in initial conversations to arrange for procurement of fingerlings. As a result, only one can commit to supplying fingerlings in advance, and only 60% of the expected need for the first year of operations.

Before obtaining permission to drill the production water well, a number of permissions from regional and state groundwater agencies are required. Other components of the facility will be delayed as a result of this. Lease payments and monthly capital expenditures and interest notes must be made. There are just a few spaces appropriate for feed storage in the warehouse facility that will be used for the project. The requirement to acquire feed in large loads adds to the complexity. Condensation and humidity cause quality issues in stored feed, leading to the need to remove feed before it can be used. While the feed storage space is adapted with proper heating, ventilation, and air-conditioning (HVAC) equipment, temporary arrangements must be made to acquire feed in smaller loads from alternate providers.

**Figure 3.8.** Bio-filtration.
Source: https://astfilters.com/aquatic-systems/wp-content/uploads/sites/3/2016/06/Broodstock-image-Paul-1.jpg.

## 3.8. BIO-FILTRATION

Bio-filtration (Figure 3.8) issues arise as a result of the prototype systems' inability to maintain sufficient flow rates when scaled up. To achieve efficient nitrification, custom-fabricated sand filters must eventually be fitted. The standard agroindustrial project, which includes most urban aquaculture businesses, is divided into four stages: opportunity identification, analysis, and design, implementation, and evaluation. When it comes to urban aquaculture, there has never been a shortage of opportunity identification, and there have been several examples of attempted implementation. Due to the low success rate of urban aquaculture firms, there have been significantly fewer opportunities for ongoing evaluation. This, in turn, represents errors in most analytical and design phases. An urban aquaculture facility's procurement, transformation, and marketing components must be thoroughly examined and integrated ahead of time. Hopefully, putting more attention on these components of project creation can improve the success percentage of such businesses in the next few years. Producers can go to and from the city on a daily basis, and farms can supply products to city dwellers. Urban aquatic production systems may be located far from urban centers, but they are inextricably linked and interdependent due to water and nutrient movements. As a result, the notion of urban aquaculture is inextricably linked to the form of urbanization that happens in and around cities. Road development determines these patterns. Many Southeast Asian countries are seeking to build provincial centers away from their major cities, and they are often areas of aquaculture development (Tripathi & Tripathi, 2011).

## 3.9. URBAN AQUACULTURE

Urban aquaculture often has many links with rural aquaculture, particularly in terms of product and knowledge transfer, and as infrastructure and linkages improve, there may be a trend toward displacement from urban to rural (Figure 3.9). Because of these tight ties, clear divisions between rural and urban livelihood strategies are useless in comprehending livelihood strategies. In general, extensive systems refer to aquaculture that relies on natural food, semi-intensive systems, which cover a wide range of systems and production levels, rely on managed nutrient inputs to supplement background levels; and intensive systems, which cover a wide range of systems and production levels, rely on managed nutrient inputs to supplement background levels. Intensive culture systems, which often have a high density of organisms, rely primarily on imported nutrients, which are normally sold as commercial feeds. Culture and the exploitation of unmanaged stocks have a significant overlap. These stocks may be derived from plant or animal origin, and practices may change over time as new fisheries and livelihood possibilities become available. The high cost of land in cities compared to rural areas is a prominent element, making food production an improbable first investment decision. Food production could also be considered a transitional land and water usage. Land and water bodies set aside for building by the government or the private sector can be utilized for aquaculture informally for a limited time.

**Figure 3.9.** Fish farm in urban areas.

*Source: https://d178ivhysawugh.cloudfront.net/media/l/2400x1260/1519469733/7823–14c8692a91f5b7a36f5304e4d15b27bc-bruceswiftandwatercress.jpg.*

Informal rentals, leases, inheritance, and traditional rights may all be used to limit access. As a result, households managing patches of aquatic vegetables on the outskirts of a municipal urban lake will approach management decisions differently than those cultivating the same crop in a tiny pond on their own property. Cities and their surrounds are generally rich in undeveloped patches due to urban authorities' attempts to limit the sort of development through zoning or other mechanisms, which, along with speculative land ownership, means that cities and their environs are often rich in undeveloped patches. Many of them are utilized for aquatic food production in flood-prone Asia. Although most city governments lack the resources to do this type of analysis, the importance of urban aquaculture in lowering organic wastes that would otherwise result in higher expenditures and public health problems is rarely acknowledged.

Clearly, much of the production from aquatic systems takes place on shared property or in a shared pool. Any action that raises yields sustainably is the primary characteristic that distinguishes farming from collection. This can be accomplished by stock ownership or restrictions on access to and benefits from such stocks.

Uncertain tenure, complex access rights, and weak planning and enforcement may foster urban aquaculture. The poor, particularly migrants seeking low-cost housing, are likely to be drawn to metropolitan areas by the opportunistic use of water and wastes, as well as a place to reside.

The tendency for informal housing to grow around water sources, which is not always transient, is widely documented, as is the link between water and sanitation and multipurpose use. Water bodies therefore provide a source of secondary income, sustenance, and work for migrants in numerous places. Processing and commerce to production are now included in modern conceptions of urban agriculture.

Value addition is an essential notion for many aquatic products; thus, this commodity approach makes sense. Trading and distributing urban garbage to rural aquaculture is a widespread and tangible link between urban centers and aquaculture in outlying areas that benefits people in a number of ways, particularly the poor.

Although the returns to labor for producing and trading aquatic products may be poor when compared to other employment options in cities, urban aquaculture is likely to enhance the number and size of niches for marginalized populations to find work.

**Figure 3.10.** Catfish.

*Source: https://www.agrifarming.in/wp-content/uploads/2015/09/Catfish-Growing.jpg.*

Changes and disruptions in the provision of fertilizer inputs might contribute to the damage of urban aquaculture as cities grow. The inability to access sufficient wastewater was a serious issue for fish farmers, forcing them to switch to more expensive inputs like feeds and modifying the ratio of species supplied. Reduced availability to nutrients could be connected to the development of improved sanitation infrastructure or a lack of maintenance of existing systems. Catfish (Figure 3.10) farmers in peri-urban areas are also leaving. As the characteristics of urban regions and production technologies change, so do the opportunities for urban aquaculture. It's possible that a large amount of traditional aquaculture took place in cities.

**Figure 3.11.** Rural aquaculture.

*Source: https://www.researchgate.net/profile/Jonathan-Munguti-2/publication/311445554/Figure/fig3/AS:436292950466562@1481031753129/Aquaculture-transforming-life-in-rural-areas-of-central-Kenya-Highland-green-algae-Farm.png.*

Rural aquaculture (Figure 3.11) on the other hand, is still hampered by these characteristics in many situations. Furthermore, rural people's propensity for catching or purchasing wild fish when it is abundant may hinder aquaculture investment. Demand in urban markets is more consistent, and aquaculture has evolved to satisfy that demand in the past. Recent research reveals that the same is true in Africa's embryonic fish culture. As a result of poorly managed cities with weak administration and/or enforcement of change and little investment in new infrastructure, the opportunistic use of urban water bodies for food production will continue.

**Figure 3.12.** Recreational fishing.

*Source: https://newsaf.cgtn.com/news/2021–09–27/Egypt-The-aquaculture-research-lab-boosting-fish-production-13SHaOFaKfC/img/997f8c9ccfa54b5c89ac819573b2dafd/997f8c9ccfa54b5c89ac819573b2dafd.png.*

Recreational fishing's (Figure 3.12) rise can be linked to any or all of the opportunities mentioned as encouraging urban aquaculture. It has evolved as a leisure interest similar to developed economies in many situations. Fee fishing operations remain the primary market for farmed tilapias in Brazil, and the industry has grown around them (Shafer, Inglis, & Martin, 2010). As long as the current trend of rural people seeking a better life in the city continues, there will be opportunities for urban aquaculture. Urbanization and industrialization are two terms that are used interchangeably. When industrialization and urbanization occur at the same time and in the same place, aquatic food production is generally boosted. Urban regions tend to have the most reliable, if not the only, infrastructure that is favorable to industrial expansion. Rapid population growth and the higher spending power of urban residents compared to rural residents result in a wide spectrum of nutrient-rich wastes. Demand is stimulated by high levels of rural–urban mobility among fish consumers. As a by-product of feeding a workforce and processing animal and human diets, industry concentrates nutrients of low opportunity value. This concentration makes garbage reuse more cost-effective and appealing to entrepreneurs who often develop profitable local systems. Increased consumption of livestock products often leads to increased poultry and pig production in urban areas, which has become a vital support for urban aquaculture.

Aquaculture may have started in cities as a method of confining wild fish in cages and tanks before selling them, and hence the techniques required to intensify cultivation may already be in place. In contrast to extensification, which can be seen around some metropolitan centers, intensification is a common response to encroaching urbanization. Given changing costs and possibilities, intensification may be unavoidable. Higher returns are required when land costs rise, and closeness to alternative resources and markets encourages innovation. The intensification of sewage-fed systems is allowing production to expand despite the loss of land and poor sewage supply. Improved management and the introduction of species like tilapias and silver carp, appear to be factors in the intensification. A steady demand for small, live fish has also aided in the expansion of manufacturing. Raising small fish that are affordable to low-income consumers and can be readily and quickly transported live decreases risk and expands opportunities in the production chain. Continuous harvesting and stocking enhance cash flow and stock visibility, as well as lowering the risk of pollution. A common development path for aquaculture firms in urbanizing areas is from food fish to fish seed and finally to ornamental fish production. It also illustrates how competition, expertise, and possibilities can promote the value chain, in addition to increasing the cost of critical resources as urbanization happens. Moving to juvenile food fish production, which has faster production cycles and better cash flow, is a common intermediate step that supports a variety of additional niches, including brood fish and live feed collection or production, as well as hatchery suppliers.

In some regions of Asia, the use of excreta and wastewater in aquaculture is common, although it is unusual in other parts of the world. However, national or local governments have only recently begun to implement planned reuse in aquaculture utilizing formally built and engineered wastewater treatment and reuse systems. Most developing countries' cities have minimal wastewater treatment, and two-thirds of urban wastewater is not treated at all. It is generally released into the nearest surface water source. Since the expense of installing conventional mechanical wastewater treatment is high-priced. Even when the latter is implemented, it rarely works as well as it could due to high initial expenditures, high maintenance expenses, and the requirement for specialized equipment and experienced specialists. The development of sustainable wastewater treatment technologies within the economic and technological capacities of developing nations is necessary to protect public health. The popularity of low-cost waste stabilization ponds with high pathogen removal rates is expanding. A complete approach based on natural treatment and reuse of nutrients and water that is both energy and cost effective, provides employment and low-cost food for the poor, and provides financial encouragement to stimulate effective operation and maintenance of wastewater treatment facilities is required.

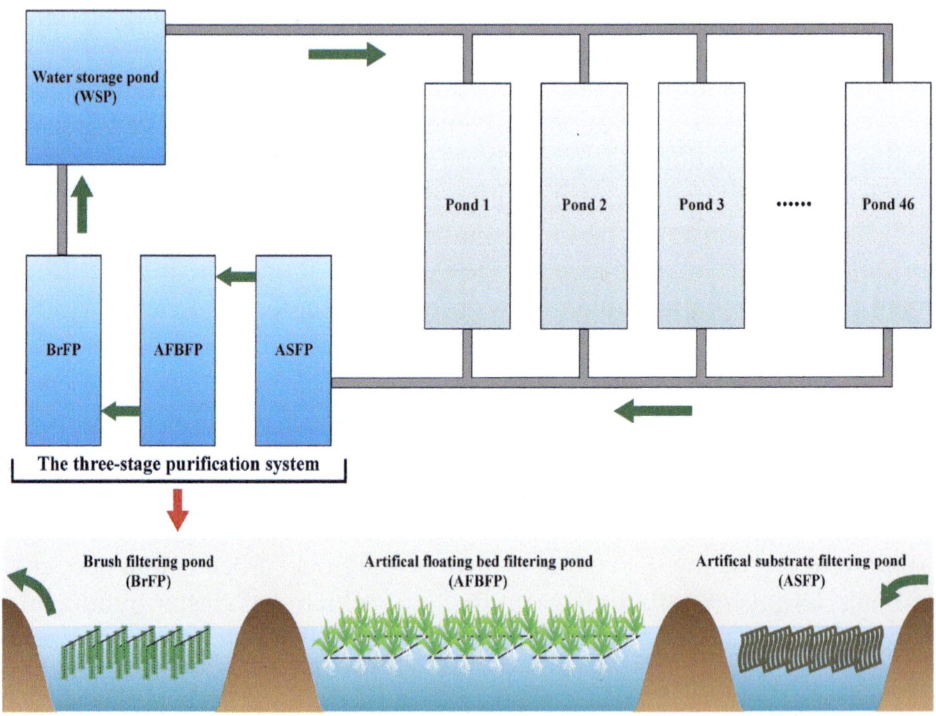


**Figure 3.13.** Wastewater treatment.

*Source: https://www.frontiersin.org/files/Articles/524568/fmicb-11-00930-HTML-r1/image_m/fmicb-11-00930-g001.jpg.*

## 3.10. WASTEWATER TREATMENT

Wastewater treatment (Figure 3.13) and reuse schemes based on aquaculture could play a significant part in urban poverty reduction efforts. Because of the rapidly increasing scarcity of renewable water resources, particularly in semi-arid and arid regions, there is significant interest in using treated effluents in many parts of the world. Wastewater is a broad term that encompasses a wide range of human excreta applications in aquaculture (Chatla, Padmavathi, & Srinu, 2020). Because of an absence of or inadequate sanitation technologies, contaminated surface water refers to natural water bodies like rivers and lakes that contain human feces.

Domestic sewage is the water-borne excreta from flushed toilets, to which the term wastewater is frequently applied, and it can contain varying levels of industrial wastewater or sewage. If the sewerage system is linked to storm drains, sewage may also contain urban run-off. Black soil can be utilized directly in grow out systems via a fishpond system, eliminating the need for transportation because the use is on-site. In contrast to direct use, when wastewater is delivered, gravity flow, or pumping through a reticulated network of sewage pipes or channels, indirect use involves the use of contaminated surface water. Fecal solids can be manually delivered in containers by foot, bicycle, or motorcycle, as well as by boat or vacuum truck. The intended use of wastewater from a receiving water body, such as a stream draining an urban area, is known as indirect usage.

On-site treatment and off-site treatment of waste are both considered low-cost treatments. Mechanical treatment or biological treatment in stabilization ponds are all options for traditional sewage treatment. Wastewater is not directly used in aquaculture. Though nutrients are taken up directly by aquatic plants, wastewater nutrients are used as fertilizer to produce natural food for fish such as plankton or benthos. In order to reduce the amount of land allocated to treatment operations, high organic matter loading rates are employed to treat wastewater in stabilization ponds.

Because wastewater-fed aquaculture is common in locations where freshwater is scarce, ponds for fish culture are first filled with the only source of water available, usually contaminated surface water or raw sewage containing a high percentage of organic matter. The anaerobic water filling the pond is left to stand for 10–20 days in this first step of the process, known as primary fertilization, until natural biological purification processes have reduced the BOD and increased dissolved oxygen to a sufficiently high concentration for fish to be stocked. In a process known as secondary fertilization, subsequent additions of anaerobic water, intermittently in small ponds to more or less continually in large ponds, maintain pond fertility to supply ample natural food for fish.

## 3.11. LARGE SCALE SYSTEMS

Large-scale systems are almost always for commercial purposes. A large variety of hypothetical or experimental systems have been proposed. The bigger system in which the aquaculture system physically operates is called a resource system, and it is usually built on agricultural or waste land, or in a wetland or water body (Blancheton, Attramadal, Michaud, d'Orbcastel, & Vadstein, 2013). Fish or aquatic plants can be grown in wastewater. Aquatic plants can be farmed in open water, such as fecal contaminated rivers, lakes, or reservoirs, or in ponds, the latter being the most common type of fish culture facility. In open water bodies, fish can be kept in cages. In a polyculture, fish culture in ponds can be combined with aquatic plants. Pond water from the aquaculture facility can be used to irrigate crops on the dyke, most commonly vegetables. Irrigation of crops, field crops, fruit trees, rice, and vegetables are also possible with effluent. Gravity flow, pumping, and sprinklers are all ways for wastewater to enter the system. Although inadequate sanitation, as in many developing countries, leads to enhanced output of both capture fisheries and aquaculture through eutrophication of surface water, wastewater usage in aquaculture is not a traditional practice in Bangladesh.

The majority of wastewater reuse in aquaculture is vulnerable or declining, with few new schemes being deployed. The goal of international-funded research and development projects in Peru was to create a wastewater-fed aquaculture system that could be used in other Latin American countries. Attempts to spread it to other cities in the region, however, have been unsuccessful. Rapid urbanization, with its high opportunity cost and thus limited availability of peri-urban land for relatively land-intensive reuse schemes; increasing industrialization, with the mixing of domestic and industrial wastewater with toxic chemicals; and rising living standards, with more affluent urban consumers preferring higher-value carnivorous and pellet-fed freshwater and marine fish over relatively low-value herbivorous freshwater and marine fish. Another key barrier is most planners and engineers' lack of understanding of aquaculture as a very low-cost technical option for wastewater treatment and reuse. Mechanical treatment devices use far less space than stabilization ponds, but due to high operating and maintenance expenses, they frequently fail to function correctly in tropical poor countries. As seen by the recent introduction of projects in Bangladesh and India, wastewater-fed aquaculture has the potential to provide employment and low-cost fish for the poor in peri-urban areas of developing countries. It could be especially useful in dry and semi-arid areas, where the need to reuse wastewater for both water and nutrients is growing.

# CHAPTER 4

# Fish Quality and Safety

CONTENTS

## 4.1. INTRODUCTION

Most consumers believe seafood is healthy and nutritious, there is a lack of a complete knowledge of the factors that influence seafood intake, such as consumer motivations, barriers, quality perceptions, and information needs. Although it is commonly recognized that eating seafood, particularly omega 3 fatty acids, lowers the risk of cardiovascular disease, less is known about the additional health advantages of eating seafood. Some people are concerned about the safety of seafood since some items may contain toxins that cause disease or components that may have long-term negative health impacts. As a result, it's critical to remove these roadblocks so that consumers can reap the full benefits of seafood. Because of uncertainties about seafood safety, the presence of environmental pollutants in farmed fish has gotten a lot of attention in recent years (Figure 4.1). Because the majority of food safety concerns in farmed fish have been related to fish feed. Persistent organic pollutants (POPs) and metals are the two main types of environmental contaminants (Jacobs, Covaci, & Schepens, 2002).

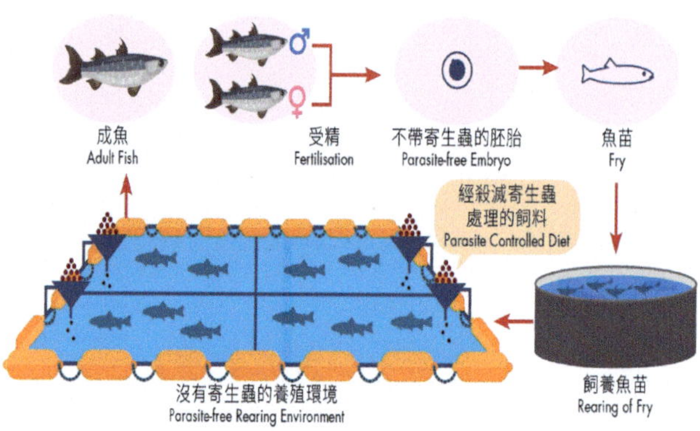


**Figure 4.1.** Fish food safety.

*Source: https://www.cfs.gov.hk/english/multimedia/multimedia_pub/images/multimedia_pub_fsf_148/fsf_148_2.png.*

## 4.2. CONTAMINATION

Chemical food contaminants (Figure 4.2) are components that are not intended to be present in food and can be detrimental to consumers if consumed at high amounts. These pollutants are distinguished from other potentially dangerous chemicals in food, such as coloring agents and micronutrient additions, by their unintentional introduction and non-functionality. The term contamination does not describe the degree to which a dietary component poses a risk to human health. This will be determined by a number of factors, including the component's toxicity, its concentration in feed, its transfer from feed to edible tissue or bioavailability, and the amount of the food product consumed. Chemical pollutants can enter the food supply chain at any point. Food can be polluted during processing, packing, or farming practices like pesticide contamination of feed ingredients. Chemical contamination of wild fish and, as a result, marine fish feed ingredients and farmed fish is a cause of pollution in the aquatic environment. This is especially true for persistent pollutants that easily biomagnify in the aquatic food chain and can be detected in relatively high quantities in fish species captured for industrial uses like the manufacturing of fish feed ingredients (Rosa et al., 2020).

**Figure 4.2.** Fish contamination.

*Source: https://sentientmedia.org/dev/wp-content/uploads/2019/04/fish-farming-cruelty.jpg.*

## 4.3. BIOACCUMULATION

Bioaccumulation (Figure 4.3) is the transfer of contaminants from all channels of exposure, including water, sediment, and food. The transmission of pollutants from the water to an organism is known as bioconcentration. The term biomagnification refers to the increase in the concentration of a pollutant from one trophic level to the next.

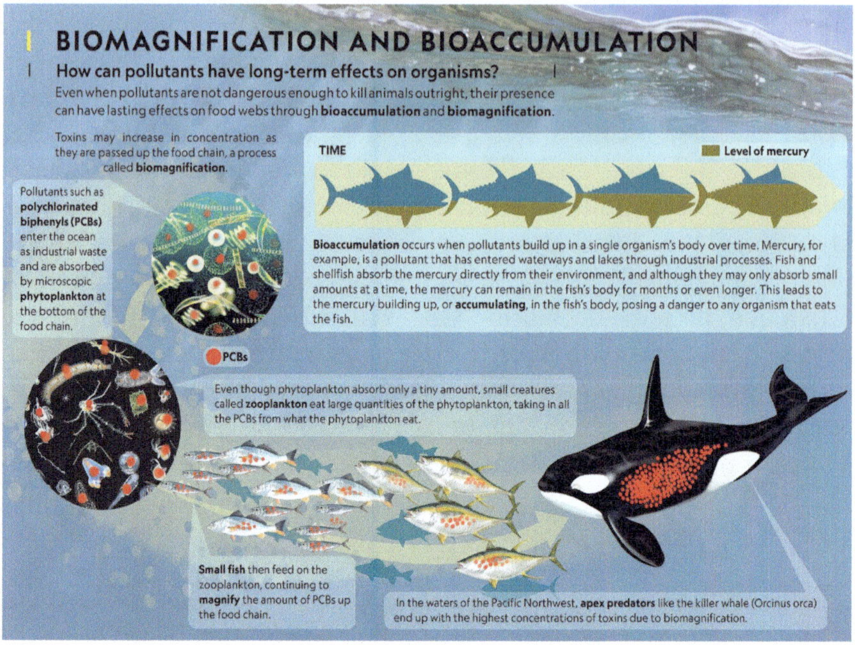


**Figure 4.3.** Bioaccumulation.

*Source: https://media.nationalgeographic.org/assets/photos/931 /115/460ca1b 4-b420–4b5e-8887–95e65f2be650.jpg.*

Dioxins and polychlorinated biphenyls (PCBs) are persistent lipid soluble pollutants that rapidly bioaccumulate in fish fatty tissue and biomagnify up the food chain. Seasonal variation, regional distribution, trophic position, age, and fish size all influence the concentration of pollutants in wild fish. Not all of the potentially harmful components found in fish are the product of pollution. When compared to the terrestrial environment, the aquatic ecosystem has comparatively high levels of some elements like mercury. Compounds with a carbon structure are referred to as organic pollutants. The majority of these components were or are currently man-made, produced by industry for a variety of purposes. Some are created by natural processes such as forest fires or volcanic activity, or as by-products of industrial activities. These POPs are often fat soluble, semi-volatile, and some are potentially hazardous at low doses, in addition to being persistent. POPs are widely disseminated and easily accumulate along the aquatic food chain, and they're frequently discovered in high concentrations in oily fish species and fish oils. Dioxins, PCBs, brominated flame retardants, and organochlorine insecticides are harmful (Chen et al., 2018).

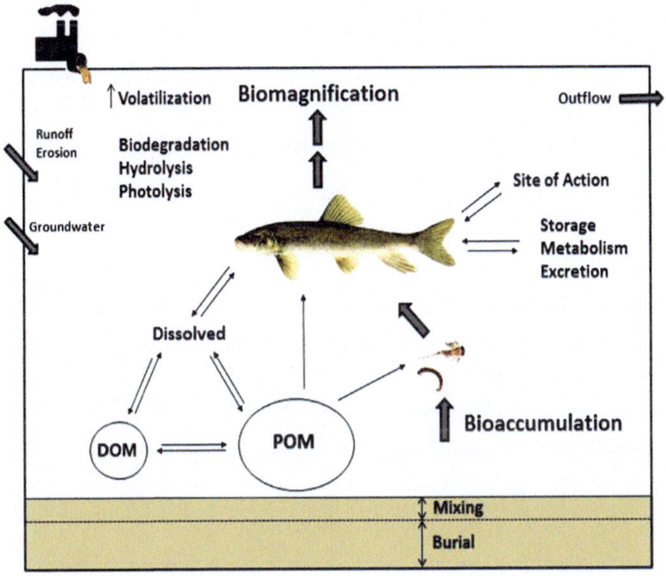


**Figure 4.4.** Polychlorinated dibenzo-para dioxins.

*Source: https://d3i71xaburhd42.cloudfront.net/75ee28bae8bdb6fd12bfb5dadb b4b5b922ad8001/19-Figure1.2–1.png.*

## 4.4. HYDROCARBONS

Polychlorinated dibenzo-para dioxins (PCDD) (Figure 4.4) and polychlorinated dibenzofurans (PCDF) are two categories of chlorinated hydrocarbons that are referred to as dioxins. The core structure of these groups is quite similar, with 210 distinct forms depending on the position or number of chlorine atoms. When it comes to the toxicity and durability of certain forms, the position of chlorine atoms is crucial. Toxic equivalency factors (TEF) have been assigned to the various forms, expressing their toxicity in comparison to 2,3,7,8 TCDD. The concentration of dioxins is measured in toxic equivalents (TEQ), which are calculated by multiplying the concentration of individual forms by a toxic equivalency factor (TEF) and summing the results to get the total TEQ in a mixture of the 17 congeners (Amlund, Berntssen, Lunestad, & Lundebye, 2012). The TEFs developed by the World Health Organization (WHO) in 1997 have been the most commonly utilized by government entities for human risk assessment. Dioxins are undesirable by-products of a variety of industrial processes, including the production of magnesium, aluminum, and nickel, paper making, oil refining, and the production of industrial and agricultural chemicals. They can also come from high-temperature combustion operations like chemical and municipal incinerators. When hydrocarbon and chlorine mixes are subjected to high temperatures, dioxins are formed. They are long-lasting compounds that are fat soluble. Dioxins are found in most foods, especially fat-rich foods like milk, meat, and fish, and are widely disseminated in the environment.

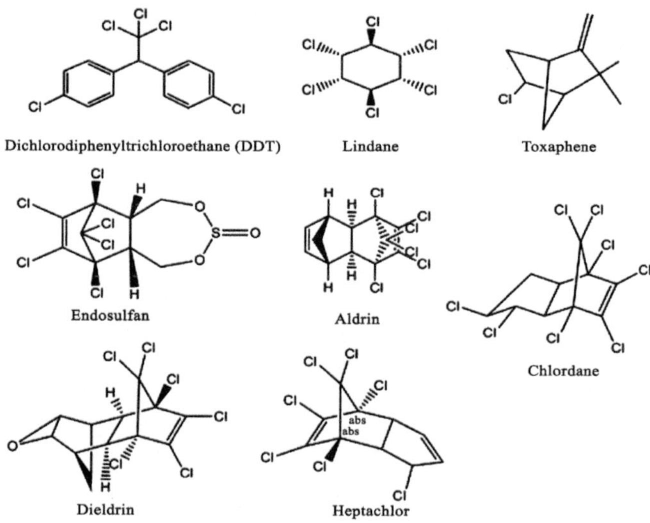


**Figure 4.5.** Organochlorine chemicals.

*Source: https://ars.els-cdn.com/content/image/3-s2.0-B9780128193044000038 -f03–01–9780128193044.jpg.*

Organochlorine chemicals (Figure 4.5) like dioxins, are polychlorinated biphenyls (PCBs). They are made up of two phenyl rings joined by a carbon link. There are a total of 209 distinct compounds that can exist depending on the quantity and position of chlorine atoms (Masci, Orban, & Nevigato, 2014). PCBs have a long history of application in complicated combinations with a variety of technical features. Some PCB congeners have a structure that is extremely similar to dioxins and have biological effects that are similar to dioxins. Dioxin-like PCBs (DL-PCB) are a group of 12 PCB congeners that have been given toxic equivalent factors (TEF) to represent their potential toxicity when compared to the most toxic dioxin. There are four non-ortho PCBs and eight mono-ortho PCBs among the 12 DL-PCB. Non-ortho DL-

PCB have chlorine atoms in the outer ring rather than the inner ring and are most comparable to 2,3,7,8-tetra-CDD. The mono-ortho PCBs are DL-PCB that have only one chlorine atom in the inner position in addition to chlorine atoms in the outer locations. PCBs, like dioxins, are widely distributed in the environment and are highly persistent and lipid soluble. Chemical oxidation or bacterial action do not quickly break them. PCBs are mostly man-made and can be found in animals and sediment. Direct dumping of PCB-containing garbage into rivers, lakes, and coastal areas, as well as air deposition into massive water systems, have contaminated the aquatic environment. PCBs were used all around the world, especially in industrialized countries. They, unlike dioxins, were created with a specific purpose. They were widely utilized as technical mixtures, such as heat exchange and hydraulic fluids, dielectric fluids in transformers and capacitors, and in the production of paints, plastics, adhesives, and flame retardants. Chronic exposure to PCBs was discovered to be detrimental to both persons and the environment in the early 1970s.

The identification of the PCB congeners that contribute the most to PCB combination toxicity has led to a focus on those that elicit dioxin-like responses. The DL-PCB bind to Ah-receptors and are poisonous in the same way that is in 2,3,7,8-TCDD. PCBs that aren't dioxin-like bind to the Ah-receptor poorly and thus don't cause dioxin-like toxicity. Because both dioxin-like and non-DL-PCB are frequently included in PCB combinations, it is difficult to discern between their effects. Furthermore, technical PCB mixes may contain dioxins or dioxin-like components in trace concentrations. Technical PCB combinations have toxicological effects similar to dioxins, including effects on the liver, thyroid, immunological function, reproduction, and behavior, and carcinogenicity. Individual exposures to non-dioxin-like PCB congeners have been linked to changes in thyroid, liver, and brain biochemistry, as well as estrogenicity, immunotoxicity, reproduction, and neurodevelopment. In the EU, maximum limits for dioxin-like PCB in feed and food have been established.

## 4.5. PCB

PCBs, on the other hand, are persistent and fat soluble, and they appear to accumulate in the aquatic food chain. The PentaBDE, OctaBDE, and DecaPBDE combinations are three commercially accessible PBDE congener mixtures. Several congener groupings are frequently found in commercial combinations. Lower brominated PBDE congeners are stable in the environment and accumulate in commercial Penta and octa mixtures. Higher brominated PBDE congeners are less persistent and can be broken down to lower brominated congeners, which are employed in octa-mixtures. The DecaPBDE combination consists primarily of the deca-BDE congener (PBDE-209), which is not environmentally stable and is also debrominated to lesser brominated congeners. Tri-BDE-28, tetra-BDE-47, penta-BDE-99, penta-BDE-100, hexa-BDE-153, hexa-BDE-154, and hepta-BDE-183 are some of the most common PBDE congeners identified in marine feed ingredients and farmed fish. Brominated flame retardants are chemicals that are manufactured in large quantities to lessen the risk of fire and its spread. Brominated flame retardants can be found in a wide range of products, from fabrics with polyurethane foams to plastic in TVs, computers, autos, and construction materials. Despite the fact that BFRs are considered novel pollutants, they have been discovered in fish since 1981. Between 1970 and 1998, the amount of PBDE detected in human breast milk increased, raising concerns about these POPs being an environmental threat to human health. BFRs can be spread by food derived from animals. As with PCBs, foods with a high fat content, such as oily fish, are the main sources of dietary exposure. Inhalation of indoor air and skin contact are two more probable pathways. PBDE is not regarded as genotoxic or carcinogenic to humans; it also does not have Ah-like qualities, suggesting that previously described Ah-mediated effects could be due to dioxin-like impurities. Thyroid hormones are affected by PBDE, which also produces hepatic and thyroidal histological abnormalities, as well as neurotoxic and behavioral consequences. The capacity of PBDEs to alter the thyroid hormones T3 and T4 appears to be their major non-lethal toxic effect. In most species, these hormones play a crucial role in controlling metabolism and development. There is currently no international law for acceptable amounts of BFR in food, while there are tolerated daily intakes (TDI) for some BFR in the United Kingdom.

## 4.6. PESTICIDES

Pesticides are a class of synthetic chemicals created solely for controlling undesirable organisms that can harm human health or cause crop damage. Some of the most well-known environmental contaminants of aquatic systems are organochlorines, which include DDT, aldrin/dieldrin, endrin, toxaphene, chlordane, heptachlor, hexachlorocyclohexane etc. These pesticides are durable and lipophilic, therefore they tend to biomagnify in the same way as other organochlorines, such as PCBs. A number of these chemicals have also been demonstrated to pose a health risk to humans. Organophosphates and carbamates, for example, are easily metabolized or excreted and do not biomagnify in the marine food chain. Persistent pesticides, such as DDT, have become prevalent environmental toxins as a result of their widespread usage over several decades. Several organochlorine pesticides have been phased out in most parts of the world, yet they can still be found in aquatic biota. Insecticides, fungicides, plants, snails, and rodenticides are all developed to manage specific organisms. Pesticides, on the other hand, have been demonstrated to have negative impacts on non-target organisms such as fish and humans, as well as wild animals (Vogt, 1987). The non-specific disturbance of the central nervous system (CNS) is their principal harmful effect on target animals. Organo-chlorine pesticides are the most important group of pesticides in terms of background contamination of marine feed ingredients, as well as the most relevant in terms of farmed fish safety. Humans may experience a variety of negative consequences after consuming environmentally contaminated seafood. Hormone disruption, unfavorable reproductive consequences, immunotoxicity, neurological effects, and carcinogenic effects are among them. Several persistent organochlorine pesticides that bioaccumulate in the food chain are subject to EU legislation in feed and feed additives.

The majority of metals are found in the aquatic environment as a result of geological processes such as rock weathering, ore production, and volcanic activity. High background levels of some metals can be found in geographical areas with special geological activity. Metals can enter the environment from a variety of sources, including natural sources and human activities. Metals can exist in both biological and inorganic forms. Organically bound metals are primarily found in the aquatic environment and are not found in large quantities on land. Metals' chemical form, or speciation, is essential in terms of their potential toxicity and carry-over from feed to farmed fish. Metals can also be found in feed ingredients, which can be a significant source of metals in farmed fish. Sulfur, phosphate, and carbonate structures in proteins are all readily attached to most of these metals. Metals, unlike lipophilic POPs, are frequently linked with the fish protein fraction. The main dietary source of metals in farmed fish is the marine protein part of fish feed; nonetheless, certain organic forms of the metalloid arsenic are present in fish oils.

## 4.7. MERCURY

Mercury is present in the natural environment in a variety of chemical forms, which can be generally classified into the groups. Bacterial activities in marine and freshwater habitats can transform inorganic mercury molecules into organic mercury, such as methylmercury. Organic forms of mercury are the most poisonous, with methylmercury being the most common in fish, whilst inorganic forms are less harmful. Methylmercury builds up in organisms and is biomagnified as it moves up the food chain, resulting in high amounts in marine predators. Mercury levels in fish are also affected by their size and age (Figure 4.6). Despite the fact that mercury is a naturally occurring element in the environment, anthropogenic contamination is common. Mercury is created by mining and smelting ores, and it is used to make chlorine and sodium hydroxide, as well as thermometers, batteries, and dental fillings. It is also employed as a catalyst in chemical processes, as well as in some fungicides and medicines.

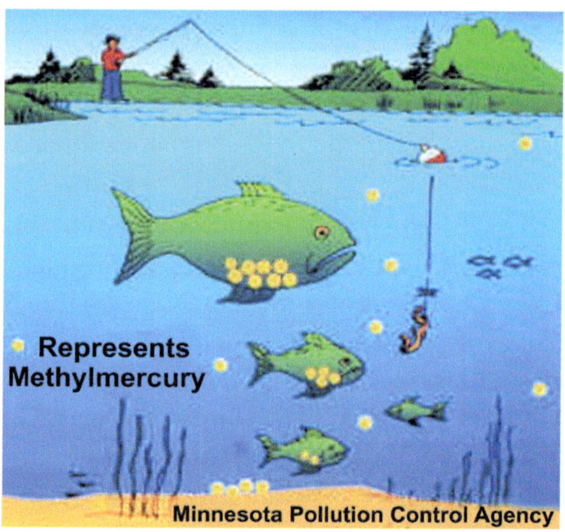


**Figure 4.6.** Mercury in aquaculture.

*Source: https://people.uwec.edu/piercech/hg/Pictures/fish.jpg.*

Maintaining ideal conditions for genetically modified crops and animals is becoming increasingly important in modern food production (Zupo et al., 2019). Best management methods include maintaining a timely and precise delivery of nutrients as well as protection from harsh growth conditions, such as pest infestations, which refers to any organism that reduces the farmed product's performance. Pesticide usage has become increasingly intensive and ubiquitous, with possible public health and environmental consequences, in almost all kinds of modern food production, including aquaculture. The desire for firm regulatory limitations on the use of chemical pesticides is being met by a demand for cheaper food. Consumer awareness has risen rapidly, fueled by concerns about long-term health effects and a competitive media, as have expectations for safe food that also include a variety of other qualities.

Pesticide usage is increasing internationally, both in terms of volume and diversity, in food production in general, as well as in broader human activities and, more specifically, in the production of farmed aquatic goods. In addition to the dangers to aquaculture from pesticides used elsewhere, the relative risks to the environment and persons associated with greater pesticide use in favor of aquaculture intensification must be evaluated. The health and productivity of the fish, as well as the effects on consumer health and attitudes, are all worthy of consideration. As pressures on wild stocks and demand from increasingly urbanized populations have increased in the last two decades, aquaculture has become increasingly important in supporting fish consumption. This firmly controlled production is based on a diverse set of species and culture systems that operate at a wide variety of intensities and productivity levels while remaining spatially constrained. As a result, four countries account for nearly all world farmed Atlantic salmon production, with China accounting for nearly 70% of global farmed output. The broader economic background is also dynamic in such fast-emerging economies, and agriculture is typically changing and intensifying, with higher pesticide use as a result. Cross contamination is more prevalent in clusters of value-added food production, such as horticulture and aquaculture, due to increased multiple uses of water within watersheds and irrigation systems.

Before looking at pesticide use trends, it's important to understand the nomenclature, classifications, and classes that define pesticides in terms of their nature, functions, and risk levels. Pesticides, like other chemical products including disinfectants, fertilizers, hormones, anesthetics, and therapeutic medications, are widely used in aquaculture, but only a handful are particularly designed for that purpose. Pesticides can be natural or synthetic, come in a variety of shapes and sizes, and have a variety of properties and modes of action. Pesticides are divided into a variety of groups based on their roles, characteristics, and toxicity. Herbicides, as opposed to insecticides and fungicides, have the highest number of active components. Pesticides are widely used in forestry, amenity use, preservation of built structures, manufactured products, homes, and gardens, food, and commodity storage, animal husbandry, public hygiene and pest control, human medicines, and general aquatic and marine pest control. Pesticides have had good impacts, such as enhanced crop and livestock yields, improved food safety, human health, quality, and duration of life, energy use, and environmental degradation, despite their well-documented negative effects. The nature and severity of their impacts are influenced by their form, characteristics, and concentrations, as well as other factors such as the nature of the environment and the sensitivity of vulnerable organisms. Pesticide hazard levels have been classified by the WHO based on its formulation and acute toxicity to rats through oral and dermal exposure.

Pesticides are used directly in aquaculture to control organisms that cause physical harm and disease to stock, compete for resources, or have a detrimental impact on the culture environment and system efficiency. Pesticides can be used in feed or as bath, dip, or flush treatments to prevent and treat damage to livestock. The pesticides used and how they are applied are determined by the target organism, the pesticide's nature, and the cultivated species. Bath, dip, and flush treatments are the most frequent techniques of controlling ectoparasites, although in-feed treatments are often the most appropriate for internal parasites, where pesticides have more direct access to the pathogens in both situations.

**Figure 4.7.** Bassia longifolia.

*Source: https://upload.wikimedia.org/wikipedia/commons/thumb/8/80/Madhuca_indica_%28Mahua%29_in_Hyderabad%2C_AP_W_IMG_0068.jpg/380px-Madhuca_indica_%28Mahua%29_in_Hyderabad%2C_AP_W_IMG_0068.jpg.*

Bath and dip treatments are best for dosage management in tanks or cages with tarps where enough oxygen levels can be maintained even when water exchange with the outside environment is limited. For brood stock that is more prone to stress during the procedure, short-term dip treatments are generally used. Flushing treatments for longer periods are particularly appropriate in raceways and flow-through systems when water flow cannot readily be decreased and oxygen levels cannot be artificially maintained. Some compounds have various effects on system management, notably in terms of maintaining acceptable water quality, in addition to their pesticidal characteristics. As a result, quicklime has long been used to disinfect earthen ponds, often in conjunction with drying. Its potent oxidizing pesticidal effects also help to maintain pH balance.

Mahua cake, a natural product (Bassia longifolia) (Figure 4.7) frequently used in India to disinfect nursery ponds, kills insects and bigger zooplankton that prey on newly released hatchlings, as well as acting as an organic fertilizer. Pesticides are also used to combat net fouling species such as algae, macrophytes, tunicates, and bivalves, which gather and limit water exchange in fish cages. Cage nets are frequently washed, dried, and dipped or soaked in viscous pesticidal solutions, many of which are copper-based, to prevent fouling. Pesticides are frequently sprayed in liquid form or placed in solid form on the pond bed after a time of drying in pond systems for prophylactic action against plant and animal competitors for resources. Although different herbicides can be used to control floating, emergent, and submergent weeds, copper sulfate is the most extensively used herbicide, notably for algal control.

Predatory and other competitive animals can be controlled with a variety of pesticides. Non-specific pesticides, such as chlorinated hydrocarbons like DDT and endrin, can be deadly to both animals and plants. To prevent predation and increase the availability of live initial feeding organisms in carp nursery ponds, a special procedure based on the use of organophosphates has been devised and widely implemented. Larger competitive zooplankton and predatory insect larvae are removed by a concentration of roughly 1 ppm of dipterex or related products, allowing smaller rotifers, which are rather tolerant to such compounds, to dominate. After a few days, the quick breakdown rates of these organophosphates allow the re-establishment of copepods and cladocerans, which subsequently become the preferred meal of the growing carp fry.

## 4.8. AQUACULTURE SYSTEMS

Through both man-made and natural processes, pesticides can infiltrate aquaculture systems and products indirectly. Spray drift, run-off, and seepage from agricultural land into aquatic systems that contain or feed freshwater or marine culture systems are some prominent examples. Pesticide penetration into aquaculture systems via groundwater reservoir irrigation and precipitation is also a concern, especially in pond and tank systems that rely on these resources in warmer climates. Pesticides can therefore come from both point and diffuse sources, influencing aquaculture directly or indirectly, depending on whether or not they come into touch with the product. Pesticide production and use trends, in terms of quantity and value, should ideally be divided by location, pesticide kind, toxicity level, and target sector. Pesticides are used in agriculture for the majority of their global trade. A growing global population and rising food demand have put more strain on natural resources, necessitating the development of more intensive and efficient food production methods. The industrialization of food production necessitated the manufacture of modern pesticides.

Pesticides can impair human health in the short and long term, depending on the chemical's nature and toxicity, its concentration, and the length and type of human exposure. Pesticides can thus have long-term or short-term impacts, with or without obvious signs. People are more aware of acute symptoms than of the risks connected with chronic poisoning, which may not be immediately apparent but have a longer-term influence. Furthermore, the nature of the risk may vary depending on the amount and kind of pesticide or its metabolite exposure. People's awareness and perception of risks, which determines their actions, are thus just as essential as physical elements in determining actual dangers. A significant level of chronic or acute poisoning might result from frequent work exposure. Consumers are at a significantly lower risk, usually due to unintentional chronic poisoning. Aside from these circumstances, pesticides have been linked to intentional poisoning and suicide, particularly among those who use them most regularly, such as farm labor. Nonetheless, there are instances where consumers, while being aware of potential health risks and having the financial means to purchase safer items, choose not to do so. This situation could occur if prospective concerns do not appear to present themselves and consumers remain skeptical, as is the case with chronic poisoning risks. The source, type, and delivery of information a consumer receives can have an impact on their risk perception. Scientifically based food safety information is frequently restricted to commercial and regulatory circles.

Consumers are frequently misinformed, with the media serving as their sole source of knowledge, where risks are frequently overstated in dramatic ways. Consumer perceptions of hazards differ depending on the nature of the danger and its link to acute disease; however, consumer trust in the institutions in place to protect their health is also important. Consumer risk perceptions are higher when they believe they have little personal control and there is a strong sense of societal responsibility and control in protecting the public. These characteristics have the potential to have a significant impact on aquaculture business, commerce and consumer health by altering consumer perceptions of risk and behavior. For example, highlighting high pesticide residues in farmed salmon may increase customer desire for wild salmon, which may offer higher health risks due to pollutants in its natural habitat. Farmworkers in many nations are unaware of pesticide exposure and protection. Detecting pesticides in aquaculture systems is as difficult as detecting pesticide chemicals themselves. The wide variety of synthetic and naturally occurring pesticide chemicals, as well as the multiple possible combinations in which they are used, makes detecting, quantifying and monitoring residue levels across crops, systems and geographical areas a difficult analytical and logistical problem. There is universal agreement that certain substances, particularly those that are more persistent, such as organochlorines, have the capacity to enter the food chain and accumulate across trophic levels through eating. Numerous investigations in ospreys, prawns and rainbow trout have produced supporting evidence. As a result, detecting and monitoring the substances used in and around farming systems has received a lot of attention. Pesticides used in agricultural systems have the potential to find their way into aquaculture systems due to the integration of farming systems in many regions of the world.

Pesticides, their residues and metabolites can be ingested by fish from a variety of sources. To begin with, pesticides may be applied to fish purposely for the aim of disease treatment, such as the use of Cypermethrin bath treatments to control sea lice in salmon cages or to cure white spots in aquarium fish. Another pathway involves the purposeful or unintentional introduction of pesticides into the fish's food, such as in-feed treatments using avermectin and emamectin benzoate or exposure to residues such as endosulfan in the raw components of the formulated feed. Pesticides can also be used to prepare ponds, such as the use of organophosphates to kill copepods and other insects before stocking fish. Herbicides can be used on the border of ponds to reduce vegetation and they can also be used to control aquatic macrophytes in the culture water. Various mitigation techniques can and have been used to reduce the need for and risks presented by pesticide interactions in aquaculture, with varying degrees of effectiveness. Various mechanical, biological, chemical and radioactive water purification along with appropriate husbandry have proved successful in minimizing the admission and propagation of infections that would otherwise only be curable with pesticides in partially or entirely recirculating aquaculture systems (RASs). Controlling infections in aquaculture systems that are more open to the outside world, such as cages or in some situations, pond-based culture, can be more difficult. Natural predators of external parasites, such as cleaner fish, can be utilized with cage-farmed salmon to manage biological pests in some cases. In addition, recent pressure on the development of less harmful compounds has resulted in lower environmental and health costs associated with pest and pathogen control in aquaculture systems. Though issues like mixture toxicity, metabolites and synergistic effects must be better understood before any of these compounds can be considered safe.

The dynamism and diversity of cultivated aquatic species encourages migration and introductions, both of which enhance pest and pathogen risk. Pesticide efficacy is often weak in aquatic systems and quarantine treatments and procedures are immature when compared to terrestrial agriculture quarantine treatments and protocols. Pesticide use in food production is often controlled, reduced or eliminated as a result of trends toward best practice and organic management in other industries. Given the open nature of many aquaculture systems, this may be challenging to achieve. As a result, on the grounds of animal welfare, producers of organic Pest problems, on the other hand, are less of a problem in aquaculture than in terrestrial agriculture systems, but this is likely due to a more recent history of intense aquaculture in most cases. A significant advantage of managing pests in aquatic farming is the ability to disrupt the cycle of pests through simple draining and drying of aquatic systems. With the growing global food demand, the intensity of food production and the usage of herbicides like glyphosate, there is a heightened awareness and concern about food cleanliness and its safety for human consumption. Consumers' expectations for healthier food, with purchase assurances and ever stronger regulatory measures governing the food industry and trade in food products, particularly those derived from aquaculture, are the result of these concerns.

The market for pesticide-free and organic products has grown and relevant certification and labeling are now linked with many products, including those generated from aquaculture, where pesticide usage is restricted. Consumer demand for healthy products and laws controlling the food business have grown faster in more developed, wealthier countries than in poorer countries, albeit the latter's rise is slowing. This appears to be related to the higher cost of producing and purchasing certified healthier foods over other foods, the cost of effective regulatory enforcement, public awareness of such health risks through education, economic development and consumers' risk value in comparison to other livelihood constraining factors. It's natural that the people are ready to pay for such products depending on how important they think avoiding these risks is to their overall well-being. The obvious result is that people in poorer, less developed countries are at a larger risk of pesticide contamination from local aquaculture products than people in developed countries. Scientific study and an improved understanding of pesticide fate and effects have resulted in advancements in the control of pesticide dangers to the environment and humans from pesticide linkages with aquaculture.

While these tools should, in theory, be beneficial to trade, the environment and human health, the standards that are established, as well as the robustness of law, regulation, monitoring and enforcement, often differ between countries. This can be demonstrated in situations where particular pesticides are banned in some nations but not in others. For example, the continuous use of tributyltin (TBT) as an anti-foulant shows that even if they are banned, they are still in use as stockpiles, are depleted or are imported illegally. When it comes to international trade, these discrepancies can lead to trade imbalances, which can have an impact on consumer health and the lives of individuals who rely on these businesses in some way. As a result, despite their benefits in terms of preserving consumers' health, they have costs in terms of other aspects of their life. Starting with the treatment of pests that impact livestock, there are a number of approaches to limit the use of toxic pesticides. Stock resistance to diseases can be improved through better brood-stock management, breeding programs, genetic engineering and nutrition advancements. In addition, creating more favorable culture conditions through oxygenation, water filtration and purification, good husbandry, and other measures to isolate stock from potential sources of contamination can reduce pathogen populations and aid stock tolerance in self-contained, flow-through or partly re-circulated aquaculture systems. Controlling diseases in aquaculture systems that are more open to the outside world, such as the cage or pond culture can be more difficult in certain circumstances. In some cases, natural parasite predators can be used instead of insecticides.

In theory, each of these mechanisms has advantages and disadvantages and many of them require knowledge and skill to implement effectively.

However, they have been shown to work best when the most appropriate methods are combined as part of a comprehensive management plan by farmers who have these characteristics and are willing and able to do so. Effective pesticide interaction control in aquaculture necessitates an understanding of the industry, the dynamics of pest problems, and the wider environmental factors. Pesticide regulation and control are carried out in a variety of ways, using incentives, with diverse results for the environment, trade and livelihoods. Other approaches, such as the use of less destructive pest control alternatives and excellent management techniques in aquaculture and other influential industries are having some success. The continuous growth of global aquaculture trade will necessitate continued improvements in traceability and producer adoption of Good Management Practices. Third-party certifiers will increasingly supervise any aquaculture firm targeting high-value markets, whether in developing or developed nations and the lack of chemical contamination will be a key factor. Setting such monitoring standards, like any other norm, will be challenging. As analytical capacity to monitor an ever-increasing number of substances improves, the rigor of standards is likely to rise. The trend toward lower detectable levels of pesticide chemicals and their metabolites as well as the expense of testing is projected to continue as demand for such analysis grows.

The requirement to lower residue levels is anticipated to induce further reductions in the relative amounts of fish-derived elements in aqua diets. A proportional spike in production of omnivorous and herbivorous species like catfish and tilapia, for which lower-cost feeds are an essential part of their comparative advantage, would accelerate this trend. Public expectations for safe food will drive these changes, as will market competition to meet that need. Investment in alternative pest management measures by both the government and the commercial sector is likely to increase, but will be constrained by conflicting needs for land, labor and water. The management of hazardous pesticide inventories and production capacity will become a major governance concern. If cost-effective alternatives are not made accessible, such products are likely to remain popular among poorer producers and markets. Several practical constraints limit the use of antibacterial drugs in aquaculture. Almost all therapies, in theory, can be given via injection, as is usual in veterinary medicine. As a result, injection is reserved for fish with a high individual value such as brood stock or ornamental fish. The parenteral approach is particularly appropriate for treating invertebrates in aquaculture such as a valuable lobster stock. Bath treatment is simple to perform with high water solubility chemicals and confined to recirculating systems or small tanks. Thus, it is mostly employed in the treatment of small fish in confined water. Bath treatment has demonstrated to be quite effective in treating fish with both systemic and skin or gill diseases. However, the capacity of cations in seawater to form complexes with antibacterial medicines like oxytetracycline (OTC) and flumequine, rendering them inactive and reducing the fish's ability to absorb them, is crucial. As a result, the dose and treatment regimen in freshwater varies significantly from those in seawater. In seawater, a greater dose and a longer treatment duration are usually required. Oral administration allows for the treatment of large numbers of adult fish with little effort and has become the preferred method of administering medication to fish. Oral administration, on the other hand, has several drawbacks. It's possible that gut absorption is insufficient, and that palatability varies by species. Because of hierarchical behavior and the fact that diseased fish generally display reduced feeding, the antibacterial drug may be distributed unevenly in the stock. As a result, therapy regimens must be planned to maximize efficacy while reducing the likelihood of pathogen resistance. In an ideal world, therapy should be initiated only once the target pathogen and its medication sensitivities have been discovered.

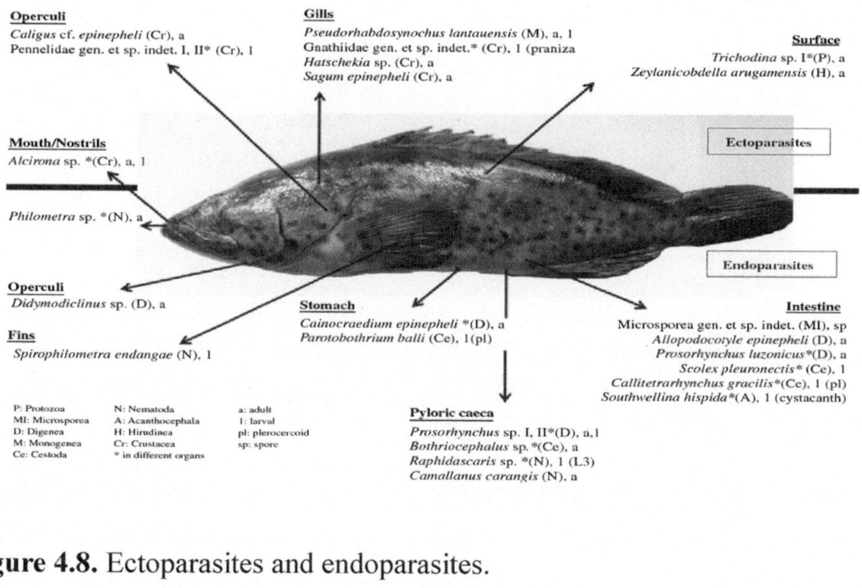


**Figure 4.8.** Ectoparasites and endoparasites.

*Source: https://www.researchgate.net/profile/Sonja-Kleinertz/publication/256480533/Figure/fig1/AS:601698733133827@1520467565124/The-occurrence-of-ectoparasites-and-endoparasites-from-the-grouper-fish-Epinephelus_W640.jpg.*

Antimicrobials cannot be used as growth promoters or in speculative chemoprophylaxis to avoid disease outbreaks in the aquatic environment, which is ideal for the development and spread of drug resistance. Medicated food pellets are used to provide the antibacterial ingredient. The medicine is either coated on the outside of the pellets or mixed in with the food ingredients prior to pelleting to make the drug a part of the pellet.

## 4.9. PARASITES

Drugs against ectoparasites and endoparasites are now accessible as medicated food pellets for oral administration (Figure 4.8). Bath therapies, on the other hand, are just as significant as orally taken medications. Previously, the most frequent way was to move the fish into a mobile pen with a tarpaulin surrounding it. The pen might be moved around a fish farm, allowing the treatment fluid to be reused. The treatment pen was oxygenated to alleviate crowding-related stress. Before releasing the chemical, the pen was dragged into open water after usage. However, using a movable pen is time consuming and has been mainly replaced with tarpaulin. The cage to be treated is brought up and surrounded with a sheet to lower the cage capacity before the allowed amount of the compound in use is put to the cage. Oxygen diffusers are installed in the cage to ensure proper oxygen levels and aid in the mixing of the compound with the water. When the weather permits, treatment cages are wrapped by a skirt rather than being entirely enclosed. The tarpaulin is removed after treatment, and the chemical is allowed to disperse into the sea. Tetracycline antibiotics were developed as a result of a systematic search for antibiotic-producing microorganisms in soil samples collected from throughout the world. Tetracycline antibiotics are currently used in both veterinary and human medicine. The first isolated component of this category, chlortetracycline, was introduced in 1948, and OTC was introduced into human medicine not long after. The actinomycetes *Streptomyces rimosus* and Streptomyces aureofaciens generate both OTC and chlortetracycline, which are antibiotics. Tetracyclines are broad-spectrum antibiotics that kill a wide range of bacteria, both gram-positive and gram-negative as well as some protozoa (Sepúlveda, Marín, & Carvajal, 2004).

Tetracyclines suppress bacterial protein synthesis, hence the bacterial ribosome is the site of action. The bacterial ribosome is smaller (70S) than the eukaryotic ribosome (80S) and it is made up of 50S and 30S subunits, whereas the eukaryotic ribosome is made up of 60S and 40S subunits. Tetracyclines have little effect on eukaryotic ribosomes because of this difference. Tetracyclines bind primarily to the 30S component of bacterial ribosomes in the bacterial cell. They appear to block aminoacyl t-RNA from reaching the acceptor site on the m-RNA-ribosome complex, preventing amino acid addition to the developing peptide chain. Shortly after its introduction into human medicine, OTC was investigated for its potential application in aquaculture. Temperature affects OTC in fish, which is normally slow and requires long withdrawal periods. Despite the fact that resistance to tetracyclines was discovered in bacteria obtained before these medicines became widely used, the number of resistant strains has increased considerably after their introduction into antimicrobial therapy. Tetracycline resistance is currently one of the most widespread among all clinically relevant bacterial species. Tetracycline-exposed bacteria frequently acquire co-resistance to other, chemically unrelated antimicrobial drugs in addition to developing resistance to the antibiotic itself.

Oxolinic acid and flumequine are the two most commonly utilized quinolones in aquaculture. Oxolinic acid and flumequine have been widely used to treat systemic bacterial infections in fish due to their low minimum inhibitory concentration (MIC) values for most susceptible fish pathogens and effective systemic distribution in fish when administered orally via medicated feed. They treat all types of bacterial infections, including furunculosis, atypical furunculosis, classical vibriosis, cold-water vibriosis, and yersiniosis. The medications are administered through medicated feed, injections in brood fish and baths in smaller fish. Because flumequine and oxolinic acid are both weak acids that tend to combine with magnesium, the pH and concentration of divalent cations in the fish's intestine play a big role in absorption. In seawater, the bioavailability of orally taken flumequine and oxolinic acid is lowered, similar to what is seen with OTC. For the same reasons, absorption from saltwater after bath treatment is significantly lower than absorption from fresh water with a good pH. To obtain therapeutic concentrations in seawater fish, high drug concentrations in the bath and prolonged exposure are required.

The bile into the intestine is a significant route for quinolones to be excreted in fish. In addition, this process is observed to be faster in seawater than in freshwater. Quinolone pharmacokinetic properties in fish have recently been examined. Quinolone resistance has become a big issue in locations where the medications have been used extensively in aquaculture. However, there is no evidence that quinolone resistance may be developed by bacterial plasmids.

Oxolinic acid was discovered in 1968 and has been shown to be efficient against gram-negative bacteria, including a variety of fish diseases. As a result, since its introduction in 1987, oxolinic acid has been widely used in Norwegian aquaculture and it has largely replaced medications such as OTC. When administered in feed to seawater fish, oxolinic acid bioavailability varies from low to moderate for both cold and temperate water species. Oxolinic acid elimination is affected by species, salinity and temperature. Antibiotics may be required in aquaculture to treat infections caused by bacteria. Such agents are being used in the cultivation of fish and other aquatic species all over the world. Acetylcholinesterase is an enzyme that breaks down acetylcholine into acetate and choline. It is linked to the membrane in both the pre and post synaptic nerve terminals. By extending the duration of acetylcholine generated at cholinergic nerve terminals, an acetylcholinesterase inhibitor indirectly offers cholinergic activity. Acetylcholine builds up in the synaptic space as a result of this. Paralysis of motor function and convulsion are two of the many symptoms of this buildup (Lai & Lin, 2009). Organophosphates bind covalently to a serine OH group in the acetylcholinesterase molecule's active site. When this happens, the enzyme becomes permanently inactive and unless a reactivator such as pralidoxime (PAM), a synthetic pyridinium compound that can reactivate inhibited acetylcholinesterase by displacing the organophosphate, is given, the activity must be restored by synthesis of new enzyme molecules. The phosphorylated enzyme will release one of the alkyl groups bound to the organophosphate if a reactivator is not used. Some organophosphates are exceedingly poisonous and were developed for military use as nerve agents. Baths are used to inject organophosphates.

The avermectins, which were first discovered from the actinomycete *Streptomyces avermitilis*, are a relatively new class of insecticides. The mechanism of action in invertebrates, as revealed in the nematode Caenorhabditis elegans, involves binding to glutamate-gated chloride channels, which results in an inflow of chloride ions and a hyperpolarized cell. Avermectins also work by raising the production of the inhibitory neurotransmitter GABA at nerve endings and extending the time that GABA binds to the receptor. Avermectins act on muscle cells and synapses in the peripheral nervous system of invertebrates, paralyzing them and finally killing them. However, in animals, the toxicity is limited because avermectins do not penetrate the blood-brain barrier and hence do not disrupt GABA-mediated neurons in the CNS. Because the blood-brain barrier in fish is not as impenetrable as it is in mammals, CNS depression and mortality have been documented in salmon treated with ivermectin at therapeutic levels. Emamectin benzoate is the avermectin used in fish farming.

A polymer is a substance made up of many tiny components called monomers that are arranged in a regular pattern. Chitin is a polymer whose monomer is D-glucose-amine and whose polymerization is catalyzed by the enzyme chitin synthetase. Chitin is a major component of insects' and crustaceans' exoskeletons. The parasite absorbs praziquantel quickly and reversibly. In vulnerable organisms, it is thought to have two immediate effects. The medication induces increased muscle activity, contraction, and paralysis at the lowest therapeutic concentration. It is most likely because of this consequence that the parasite loses its adhesion to the host tissue. Praziquantel causes vacuolization and vesiculation of the parasite tegument at a higher but still therapeutic concentration. This causes an osmotic and nutritional imbalance, as well as the discharge of the parasite's contents, as well as the activation of the host's defense mechanisms. Increased membrane permeability to some mono and divalent cations, particularly calcium, may explain the impact of praziquantel at the molecular level. A category of chemicals known as substituted benzimidazoles has comparable molecular structures. Some of these are among the most potent chemotherapeutic drugs with total larvicidal action that have ever been discovered. They do, however, have a low mammalian toxicity and do not interact with other microorganisms.

## 4.10. RESIDUES

Microbiological, immunological and chemical principles are used in the identification of drug residues in fish, shellfish and crustaceans. Microbiological methods are employed in the routine detection of antibacterial compounds and are based on the ability of these chemicals to prevent the growth of bacteria. A portion of the tissue to be investigated is injected on optimum media with various test bacteria that are specifically sensitive to relevant antibacterial drugs. After incubation, if the sample contains residues of any growth-inhibiting chemical, a zone without growth may appear. The amounts of drug residues in liver samples from the same individual are higher than in muscle samples from the same individual, hence liver samples are preferred in these tests. Although the microbiological approach may identify a wide range of antimicrobial drugs, its sensitivity is low. Microbiological findings that are positive or ambiguous should be confirmed using more sensitive and specialized chemical procedures. In the testing of drug residues, a variety of chemical techniques are available and are regularly used. Separation using liquid or gas chromatography in combination with various detectors are the analytical principles involved. Chemical approaches, in comparison to microbiological procedures, detect drug residues at considerably lower concentrations and are also more specific.

CHAPTER 5

# Bioeconomics of Indoor Fish Farming

CONTENTS

## 5.1. INTRODUCTION

The integration of biology and ecology with the economic aspects that govern fishers' behavior in location and time is required for systematic fisheries research that may create and evaluate hypotheses associating different management and regulation structures with fishery performance. Because the fundamental qualities of fish resources provide diverse contexts of human interdependence, the same institution, management, or regulatory framework can produce different results when applied to fisheries with different resource and community contexts. Human interdependencies in resource usage are directed by the degree of resource mobility in geography and time. Modern fisheries bioeconomics can help establish strategies for dealing with the complexity of overexploitation in marine fisheries, which are mostly influenced by natural fluctuations, changing coastal ecosystem dynamics and a lack of good governance. The degree to which the outputs created as a result of employing the resource for which the property right is distinct are completely under the control of the owner to utilize is referred to as exclusivity. Similarly, the extent to which the owner is responsible for all costs related with the resource's utilization (Huang, Chiou, Khac, Hsiao, & Chen, 2016).

## 5.2. RESOURCES

It also refers to the resource's ability to be used and managed without outside intervention. It's crucial because the owner's ability to claim benefits or be accountable for costs will be determined by how long they can claim benefits or be responsible for costs. This refers to the power to assign a right to another person. This is also necessary in order to ensure that the resources are put to the best possible use. Other features and attributes of fisheries resources that can affect short and long-term fishing behavior and exploitation patterns, both with and without regulation, should be considered when thinking about fisheries management and regulation. High exclusion costs, high transaction costs including information and enforcement costs, behavior and a social situation must all be taken into consideration. Just because there are regulations allowing only a certain number of people to fish does not mean that everyone else is effectively excluded. Enforcing rights and rules in maritime fisheries is not only logistically challenging but also exceedingly costly due to the mobility and migratory nature of most fish resources. To maintain participation limited to permitted individuals and fishing operations limited to those that are permitted, fisheries management and control entails substantial enforcement expenditures. The areas to be regulated for oceanic fisheries are vast and traditional patrol vessel operations are ineffective and expensive.

The substantial uncertainties inherent in natural systems, as well as a variety of other biological, social, political and economic issues, add to the complexity of fisheries management, necessitating a cautious approach to fisheries management. In general, even with stakeholders' collaboration, obtaining valuable information adequate to know what is the proper management strategy and corresponding legislation is quite expensive. Furthermore, the multidisciplinary character of the data needed to assess the condition of the fisheries and the ecosystem that supports it necessitates interdisciplinary techniques that combine ecological and social sciences. Furthermore, fisheries are frequently presented with a social trap situation, which is connected to the absence of exclusivity in the rights to free fish populations. A social trap emerges when an individual fisher's motives in the short run are incompatible with the vast results. Short-run micromotives include capturing as many fish as possible to improve individual marginal advantages and long-run intended macro results could include obtaining the highest economic yield or maximum sustainable production. Long-run results are frequently dominated by short-run marginal advantages due to uncertainty about future stock availability. A sustainable output from a fishery will tend to be attainable only when the number of fishermen is limited and they act to adopt some sort of effort limitation. When the majority of community members refuse to take voluntary collaborative action to prevent resource depletion as well as when stock availability is uncertain, this sort of individual emerges.

## 5.3. FISHERIES

A fishery can be defined as a stock or stocks of fish, as well as the businesses that have the ability to exploit them. It can be a pretty straightforward strategy in which a fleet of similar vessels from a single port exploits a single fish resource. It can also be more sophisticated, with fleets from various ports harvesting fish from numerous ecologically connected stocks using different technologies. First, the system's natural capital is the fish stock, which is assessed in terms of biomass. The ability to reproduce and give new recruits, individual growth rates, natural death rates and fishing mortality rates are all factors to consider. The production function is the link between stock size, vessel activity harvest, input costs and fish price. The fleet size will fluctuate depending on the size of net returns. The fleet will tend to grow if gross returns are larger than all production costs. The biological analysis enables a consideration of how harvest and stock size interact. However, in order to comprehend the operation of a commercial fishery, one must first comprehend the degree of effort that will be produced under specific conditions. This is what a bioeconomic model is for. Commercial fishing is for a profit business, for the most of part. It is feasible to develop a model that can help forecast levels of effort and output by introducing information on prices and costs, as well as how the profit level will vary with output.

Individual operators may own their own vessels and so have property rights in the fishery's capital, but no one may claim ownership of the fish in the water. There are no property rights to the fishery's living capital. While each participant would be better off if their boat could function in tandem with a larger fish stock, no one operator can guarantee that investments in the stock will pay off. There is no guarantee that if an operator reduces effort to allow stock size to rise, others will not take the excess fish. Individual operators prefer to evaluate solely the difference between average returns per unit of effort in this case. The basic problem with an open access fishery has been demonstrated by the above-mentioned analysis. Due to a lack of property rights, private persons acting in their own interests will be tempted to overfish the stock. The fishery will get too many inputs, resulting in a marginal benefit for the remaining units that is smaller than the marginal cost. As a result, the stock will be smaller than it should be and potential profits will be lost. These findings were proved in a static model in which prices, costs and technology were expected to remain constant. When the price of fish rises or harvesting technology improves, as it has in the past, the outlook for open access fisheries usage becomes even weaker. Anything that reduces the cost of inputs connected with the creation of effort will have long-term negative consequences. As a point of comparison, in the marketplaces for most other commodities and services, higher prices, cheaper input costs or technological advancements will be beneficial because they will result in more production. These adjustments increase earnings, which incentivizes the industry to bring in more inputs and more output will be the result. The essential aspect is that in other industries, more of all the necessary inputs can be added. In the case of fisheries, however, this is not the case. While more boats can be brought in, more productive capacity in the form of a larger fish stock cannot be added. This critical distinction will result in different outcomes.

## 5.4. REVENUE

When entire revenue equals total cost, a fishery achieves open access economic equilibrium. The preceding analysis was done using the sustainable yield curve, which took into account the impacts of effort on stock size. The purpose was to explain the distinctions between optimal fishing use and open access equilibrium. Because a fishery is rarely, if ever, in a condition of balance, the discussion is crucial. When contemplating broad picture issues, the concept of equilibrium is useful, but not so much when it comes to real policy. There will be a specific stock size and a certain quantity of potential effort at any one time. It is unlikely that the equilibrium combination will be found. Stock and effort will fluctuate over time for some reasons. When thinking about regulation, it's important to understand more than simply assessing the current situation. It's critical to understand where the fishery is headed. It's also crucial to understand how to adjust the time-path trajectory so that it points in the appropriate direction. When neither the stock size nor the intensity of effort changes, a bioeconomic equilibrium is reached. It's a biological and economic balance at the same time. The stock size will not vary because catch equals growth. Alternatively, the PEC can be viewed as a line that separates the stock-effort space into effort and stock size combinations that cause stock to increase or fall. The catch for the current level of effort will be greater than growth for the existing stock size at all locations above the line, causing stock size to fall. Alternatively, catch will be less than growth at all combinations of effort and stock size below the line, causing stock size to grow. Somewhere along the PEC, the bioeconomic equilibrium will have to be reached.

It's time to return for a more in-depth look at how to use a fishery in the most cost-effective way possible. The static analysis described is adequate to explain the basic notion and show why an open access system with no or limited property rights would typically provide incentives that lead to an inefficient mix of effort and stock size. Static analysis, on the other hand, excludes the time it takes for the fish stock to react to changes in effort. And knowing the complicated nature of the advantages and losses of various temporal harvest patterns is crucial. The size of a fish supply changes over time, both naturally and as a result of fishing mortality. The net revenues that can be earned are affected by these changes. Heavy fishing will produce profits in the short term, but at the expense of a smaller fish resource and future potential gains. Alternatively, minimizing or maintaining low fishing pressure now will allow more fish to develop and reproduce, resulting in larger future potential gains but lower current advantages from fisheries usage. This is the root of the issue when it comes to fisheries protection. It's important to weigh the benefits of catching another fish now vs letting it die in the sea. This is a lot more difficult problem than it appears at first glance. It necessitates taking into account population dynamics, harvest technology, fish product valuation, time preferences at the same time. In order to achieve dynamic optimal use of a fish stock from an economic standpoint, a long-term harvest strategy that maximizes the current net value of output must be chosen. Optimal utilization can be characterized in terms of a temporal path of effort starting with a current stock size. Both harvest and stock size time pathways will be generated by the effort time path.

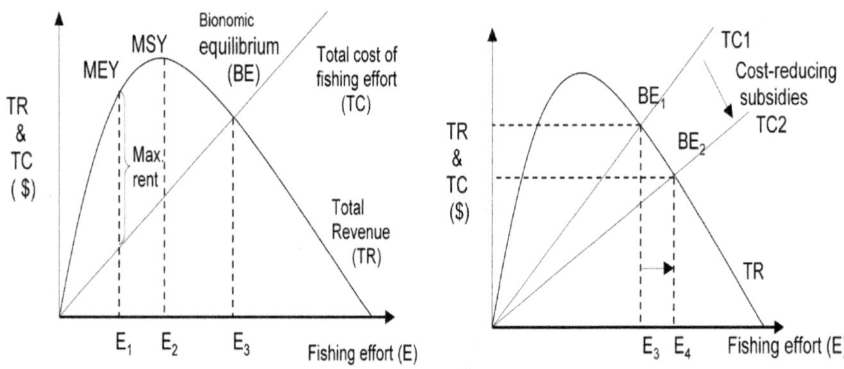


**Figure 5.1.** Schaefer logistic growth biological model.

*Source: https://www.researchgate.net/publication/278013594/Figure/fig1/AS:399632887762945@1472291313655/Static-Gordon-Schaefer-bioeconomic-model-of-fisheries-adapted-from-7_W640.jpg.*

## 5.5. MODELS

The Schaefer logistic growth biological model (Figure 5.1) has been used by certain scholars to describe the underlying theory of fisheries bioeconomics. It's a simplistic biological model, but it's nevertheless used to examine some real-world fisheries due to the relative simplicity with which the necessary data can be obtained. It's also good for analysis because it enables for simple mathematical manipulation and a smooth transition into simulation modeling. It captures the fundamentals of stock size change in the presence and absence of fishing mortality. As a result, it serves as a useful demonstration platform for the dynamics of open access, optimal and regulated consumption. The net effects of new individual recruitment as well as growth and natural mortality of existing individuals are all summarized in a simple equation in the Schaefer model. Furthermore, it is implicitly believed that natural and fishing mortality can be evaluated separately when tracking changes in stock size over time when considering fishing mortality (Bjørndal, Lane, & Weintraub, 2004). Although more difficult, utilizing an age-structured framework in which recruitment, individual growth and natural mortality are considered separately can provide a better picture of population dynamics. The economic and fishery management insights derived from the Schaefer model may largely be expressed in terms of the age-structured model. In reality, it is possible to create sustainable revenue and cost curves as well as population and economic equilibrium curves, using numerical rather than analytical methodologies (PEC and EEC). It does add value by allowing for the examination of age at first capture as well as the unique characteristics of certain stock–recruitment relationships. Predation, spawning stress, diseases, unfavorable environmental circumstances and a shortage of food and/or space are the factors that can cause natural mortality in marine animals. Most age-class models assume that natural mortality is the same for all age groups and this assumption is often set to 0.2. However, because post-larvae and juveniles are more vulnerable to predation, competition and poor environmental conditions than adults, it makes sense that natural mortality would be higher in the early stages of the life cycle.

## 5.6. CHALLENGES

We can explore the dual challenges of what stock size to target and what path to take to get from the current stock to the target stock in dynamic optimal utilization. While such questions arose from a strictly limited model aimed at maximizing the present value of net earnings, they are also at the heart of the paradigm that underpins modern-day fisheries management policy. From a conceptual standpoint, it should be possible to plot relationships that indicate how several critical factors, such as employment and the balance of payments, change in relation to stock size. This would provide a broader view of options than simply identifying where yield and net economic returns are highest. Furthermore, relative weights might be assigned to each of these variables, allowing the curves to be combined to determine the stock size at which the weighted total of benefits is maximized. However, the majority of global fishery management programs stipulate that the proper goal is to maximize long-term sustainable catch.

A harvest control rule is a procedure that determines the intended harvest limit as a function of stock size at any given time. A target framework is also used in the execution of the harvest control rule. The harvest control rule produces a limit harvest level that is designed to be the maximum amount that can be taken while still maintaining or achieving the target stock in an acceptable amount of time. The target harvest level is below the edge harvest level as a precaution to keep the stock on the targeted growth trajectory in case the estimations of current stock size, expected stock change or the edge harvest level are inaccurate while generating recommendations for the actual permissible harvest. Using objective and limit reference points, this management paradigm analyzes both stock and flow components of management. The goal and limit stock sizes serve as a benchmark for assessing the stock's current condition. The stock may be considered overfished if it is currently trading below the limit stock size. Overfishing of the stock occurs when the present harvest rate exceeds the limit harvest rate, indicating that the stock is not on the anticipated trajectory toward the goal stock size. The paradigm's practical application in modern-day fisheries management necessitates the ability to collect data and conduct research in order to conduct stock assessments that can provide current estimates of stock size and forecasts of how different harvest levels will affect short-term stock growth.

## 5.7. DECISION MAKING

Another feature of political decision-making is that the most difficult judgments are typically made during times of crisis. Management issues are almost always high on the political agenda when stocks are in crisis, because that is when drastic harvest reductions are recommended. They are frequently in dire financial problems and due to a lack of other work options are unable to accept harvest cuts and wait for the stock to recover. Because many crucial conservation decisions are not taken until the situation is already risky, political decisions are even less likely to support necessary conservation. Part of the reason for the harvest control rules and the concept of maximum stock size was to address the problems caused by the institutional framework for decision-making. Harvest control measures may appear to limit managerial input because the allowed harvest is decided using a simple rule. However, this is not the case. In fact, there exist an endless number of control rules that, if followed, will result in the desired stock size. Their shape and location will be determined by the amount of time allotted to reach the goal stock size, as well as whether the heavier load of lower harvests falls early or late in the harvest path. These are certainly policy concerns, but the key point is that these decisions be taken ahead of time (Vergara-Solana, Araneda, & Ponce-Díaz, 2019).

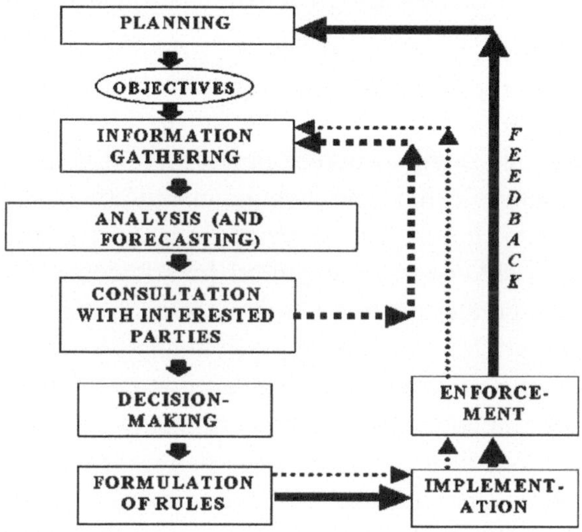


**Figure 5.2.** Fisheries management.

*Source: https://www.fao.org/3/y3427e/y3427e01.gif.*

## 5.8. MANAGEMENT

Fisheries management (Figure 5.2) which entails deciding on a target stock size and a harvest time schedule to attain or maintain it, may be a challenging and time-consuming task. However, the problem of fisheries regulation, which entails determining how to control harvest so that desired and actual catches in any given year coincide, is just as difficult. One of the issues is that, while the regulating goal can be expressed in terms of annual catch, a fishery is not a static phenomenon, as the time-path trajectories demonstrate. Things that occur in one era might have long-term consequences. This is true for regulations as well. Things done in a single period can have an impact on stock and fleet sizes, affecting the efficacy of existing restrictions and the overall ability to manage harvest in the future. In fact, it's so complicated that it's difficult to know where to start explaining it. If things were bad enough, managers would start putting restrictions on what the existing participants could do. Input restrictions such as specifying permitted types and amounts of gear and methods, as well as limiting available fishing areas or seasons, were used to limit harvest. This sort of regulation raises the cost of fishing by restricting what operators may do and creates incentives to adapt fishing tactics in order to improve catch given the limits. This has the dual impact of reducing the biological efficiency while also increasing the expense of fishing.

Managers also utilized output limitations such limiting individual fishermen's total permitted catches (TACs) or trip limits. These management tactics encourage fishermen to invent new types of gear or discover new ways that allow them to catch more fish against the rules and to do so faster than other fishermen, before any general limit is achieved. Because any fish not collected is likely to be captured by someone else, neither input nor output controls create incentives for individual fishermen to delay fish harvest. The motivation to maximize individual catches, which is part of both input and output controls, produces an uncomfortable game between managers and fishermen in which the fishermen always have the last move. Fishermen have incentives to create new fishing methods that, although legal, allow them to raise their effort while remaining compliant with the restrictions in reaction to each new measure aimed to control the amount of fishing effort they are allowed to produce. As a result of the increased harvest, management enact more restrictive measures and fishermen devise increasingly clever ways to get around them.

If the length of the boat is limited, fishermen may expand the breadth to boost fishing power. Fishermen have incentives to do everything they can to adjust inputs to capture fish quicker than their competition, rather than aiming to build boats and design equipment that can harvest efficiently. Fishermen will strive to get around the limits if input controls are employed. Because it takes time for fisherman to modify their gear or behavior, such bans can be biologically beneficial in the short term. The temporary rise in stock size, on the other hand, simply helps to fund more changes in things like boat designs that have more fishing power. When faced with a short window of opportunity to harvest, fishermen may feel driven to work in hazardous weather rather than risk losing harvests to their competitors by waiting for better weather. It's crucial to remember the aims and limitations of models as we go along, especially as we get closer to real applications. The models are built on tight assumptions that allow for a thorough examination of the fundamentals of fisheries operations. However, because numerous complicated problems are ignored, every detail cannot be captured. When interpreting the results, it's crucial to keep the stated and implicit assumptions in mind. In summary, our models can be used to derive significant inferences about the expected outcomes of various policies. Our sophisticated models make precise assumptions about the absolute and relative sizes of the economic and biological components, in addition to the usual objections about the use of models. In some circumstances, the type of result that results from the application of a particular regulation is dependent on parameter selection. Determining what to do to reach a goal harvest level in the coming season is a big aspect of practical fisheries management. This necessitates a short-run study in which the target harvest level must be evaluated in terms of existing stock and fleet conditions. To put the debate in context with the above-mentioned uncertainty analysis, it's probably best to conceive of status quo as the best estimate of the existing stock and fleet size and composition while neither is at equilibrium.

First, repurchases are only a temporary answer in the absence of a license-limitation program. Second, if the fishery's fleet is already shrinking, a repurchase may not be necessary. The fleet will be reduced to its equilibrium size if normal exit is used. The repurchase could help leaving players by speeding up the process and providing financial aid. A repurchase, on the other hand, will usually result in a dramatic jump along the trajectory, but the fleet will then resume its course to equilibrium. There will be no gain in the long run. Furthermore, a temporary reduction in fleet size is a rather crude control method. Setting it to a target stock size is really difficult. Furthermore, it is a very flexible limitation because surviving vessels will increase production of effort as stock size grows and can do so independently of stock size by altering technology. Limited-license programs, according to this analysis, can be a critical component of modern-day fisheries management regimes. They have the potential to considerably improve traditional management outcomes. They have the ability to enhance the size of the equilibrium stock and provide positive net returns. They can mitigate some of open access's dissipation issues. However, there are a few crucial limitations to keep in mind while putting them into practice. While they do not point out severe defects, they do highlight significant factors to consider when putting these programs in place.

It is crucial to remember that the models' good outcomes are contingent on the assumptions that underpin them. The models must assume that technology is at its current level. Technology, on the other hand, will evolve through time. And, more to the point, because this is an input-based program, technology is likely to evolve in such a way that the input constraints become less restrictive. In terms of the model, this suggests that any PEC shifts caused by legislation will almost certainly be followed by PEC shifts caused by technological progress. Gains in stock size and economic rewards may only be temporary. Second, the model assumes a uniform fleet and that individual vessels can be issued licenses. Of course, there are many different sorts of boats, and each will receive updates based on technological advancements on a regular basis, but in their own unique way. In other circumstances, licensing individual pieces of equipment, such as fish traps, may be more appropriate. In any scenario, determining an acceptable measure of effort is difficult. A 100-meter vessel is not the same as a 200-meter vessel, and two 100-meter vessels are unlikely to be similar. When a license must be transferred from one boat to another or a boat must be changed, there can be complications. The fundamental issue is limiting the amount of fishing power available under the limited-license program to a predetermined quantity over time.

The next phase is to choose a management measure or group of measures for attaining each aim based on the information gathered and the formulation of operational bioecological and economic objectives for the multispecies fishery. Specific management measures should be complemented with decision rules outlining how they should be implemented. Control rules specify what management action should be taken in certain situations, which are frequently defined by the value of an indicator in respect to a target or limit reference point. How the management measure will be established, what data will be collected and how data will be used to determine the measure should all be included in the decision rules. The decision rules can be developed using a variety of methods. Setting precautionary catch limitations on prey species to accommodate for predators is one method. However, because the particular types of interactions between species are rarely well understood, the degrees of uncertainty will almost certainly increase when species interactions are considered. Another technique is to generate a multispecies vector of permissible catches of target species based on observed interactions between species in multispecies fisheries, so that non-target species objectives are met. It should be noted that decision rules can be quantitative or qualitative.

The consequences of dynamic pool assumptions in overestimating stock abundance are examined, along with efforts to relax this unrealistic assumption for stationary species and various low-mobility resources through spatial modeling. Understanding the spatial behavior of species, as well as the related abundance heterogeneity in space and time and ecological interdependencies within an ecosystem framework is required for responsible spatial management and modeling of fisheries. It also necessitates a thorough grasp of the factors that influence the spatial allocation of fishing intensity. This latter point is crucial in fisheries economics, which focuses on fishermen's motivation in their fishing activity through time and space. These models are constructed on dynamic pool assumptions, which state that the resource is homogeneously distributed in space; ages are perfectly mixed and either fishing effort is applied uniformly across the range of resource distribution or the resource is able to redistribute itself. Models based on dynamic pool assumptions are insufficient for sedentary and low-mobility resources, resulting in substantial model inaccuracy. Throughout the stock range, the quality and holding capacity of local habitats varies. As a result, fishing intensity varies throughout space.

When using an MPA to spatially manage a single-stock population, it's important to evaluate the impact of situating it in relation to the fleet's point of origin. The model correctly predicted no fishing in the alternate MPA locations. Because of the high cost of steaming, effort is directed closer to the fleet's origin port in both circumstances. We have the paths of species biomass with regard to time and effort in three different scenarios under open access. They are 1. without an MPA, 2. with an MPA near port and 3. with an MPA far from port. Biomass tends to be at higher levels over time in the process of converging to bioeconomic equilibrium when the MPA is located near a port, as expected. The steaming cost of the individual fishery will determine the proportionate difference in bioeconomic equilibrium biomass for the three scenarios. A marine protected area (MPA) (Figure 5.3) is a geographical area with defined boundaries that has been declared to help conserve marine resources. Marine reserves are frequently referenced as part of a new stock and ecosystem recovery strategy. Protection of spawning biomass, providing a recruitment source for the surrounding areas, supplemental restoring of fished areas through emigration, maintenance of natural population age structure, maintenance of undisturbed habitat and insurance against management failures in fished areas are all potential benefits of establishing marine reserves (Le Gouvello et al., 2017).

MPAs assist nearby fisheries in two ways. They are net emigration of adults and juveniles from protected to unprotected areas and export of eggs and larvae that will eventually settle in unprotected habitats that are fishable. The former will be determined by resource mobility throughout space, whereas the latter will be determined by egg and/or larval dispersal processes. Within marine reserves, populations tend to grow in size, people live longer and reproductive contributions rise.

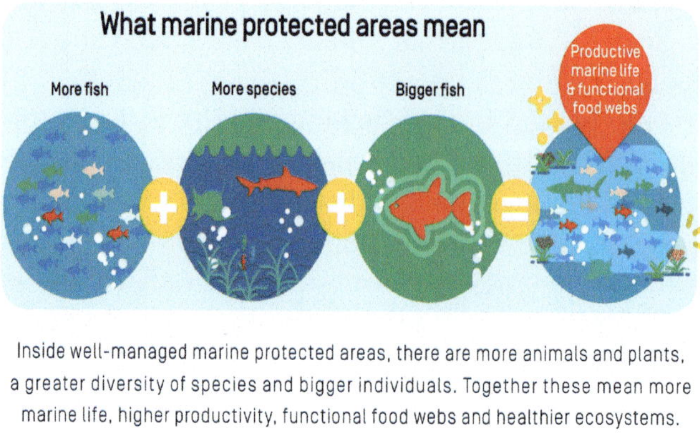


**Figure 5.3.** Marine protected area.

*Source: https://saveourseasmagazine.com/wp-content/uploads/SOSMag-Iss05-Art06-Im004-%C2%A9IvanColic-AfroGraphique-SOSFCopyright-Marine-ProtectedAreas-975x516.jpg.*

Adult biomass residing in a subarea of metapopulation distribution was used to produce recruitment in the space–time bioeconomic model. The random distribution is used to disperse recruits created by a stock–recruit relationship using this source biomass solely over the entire array the next year in both source and sink areas. The remaining biomass of mature source components is fed into a Beverton–Holt spawner–recruit model, which generates recruitment for the following year across the entire array. This seems like a reasonable paradigm to apply, especially for highly productive invertebrates and species whose optimal habitat for recruitment is limited, resulting in an effective plateau of recruit numbers.

**CHAPTER 6**

# Inland Fish Management

CONTENTS

## 6.1. INTRODUCTION

Inland fish has remained an important source of animal protein to this day, especially in locations far from the sea (Figure 6.1). There has been a quick change in attitudes toward inland water resources as a result of the strong demand for fish already in place and increased understanding of the importance of the environment to human well-being. This change is a result of a growing concern for the long-term stability of ecosystems and the ability of available freshwater resources to satisfy human needs. Humanity is shifting away from old exploitative development methods and focusing on agriculture, forestry as well as on fisheries as priorities in order to ensure the long-term viability of the world's resources so as to ensure long-term food security. Policies for conservation and sustainability are being implemented in response to the overall tightening of restrictions on natural resources. Inland waters are particularly vulnerable to these increases in resource consumption and environmental effect since they serve as collectors of all activities occurring in their basins and are among the most endangered ecosystems on the planet. Inland fisheries and aquaculture are heavily influenced by the social and geographical setting in which they are pursued and they adapt to the current world's demographic, economic and political evolution (Bonar & Hubert, 2002).

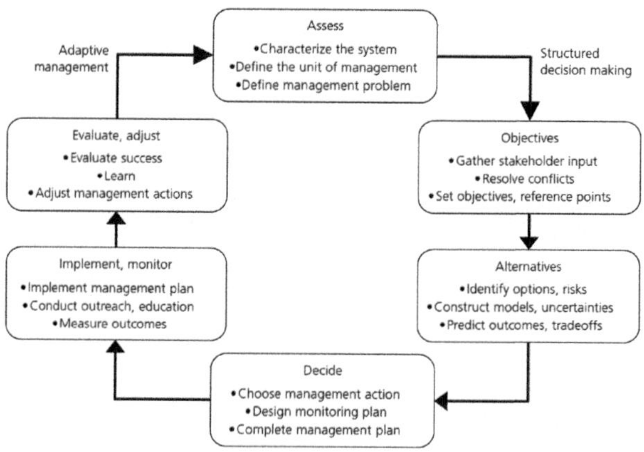


**Figure 6.1.** Inland fish management.

*Source: https://www.researchgate.net/profile/Robert-Arlinghaus/publication/283322731/Figure/fig1/AS:330416910290947@1455788938538/1-The-inland-fisheries-management-process-formulated-for-structured-decision-making-and_W640.jpg.*

Several elements can be filtered out as particularly essential in shaping future resource usage patterns, such as conditioning policies and techniques for management. Changes in human distribution have an impact on the impact of a growing population. Urbanization and demographic migrations to reside around lakes, rivers and the continent's sea margins have been key trends in recent decades. These changes result in a shift away from subsistence agriculture in rural areas, as well as a growing demand for food to be transported to urban markets via commerce. Markets in cities are more diverse than those in rural areas, necessitating a diversification of species and products. As the amount of garbage grows and other population-induced consequences concentrate more near the available water, urbanization and migration to riparian areas suggest more environmental demand on the resource.

As resources struggle to keep up with continually increasing demand, water is now one of the limiting factors to society's growth. Water supplies per capita will be significantly reduced in other parts of the world. With increased strain on a fundamental resource, it is evident that competing users of that resource, such as agriculture, home consumption, industry, transportation, power production, recreation and fisheries, will need to become more efficient and concentrated in their water use. Furthermore, the necessity to protect water quality for human use necessitates that intensive fisheries and aquaculture techniques either produce less pollution and eutrophication nutrients or be linked to downstream water treatment systems.

**Figure 6.2.** Inland fisheries.

*Source: https://images.squarespace-cdn.com/content/v1/52a9273ae4b07fa2610392dd/1564715642069-EE250OZK5PPD6UK143YM/IMG_3476.jpg.*

## 6.2. INLAND FISHERIES

Inland fisheries (Figure 6.2) in the tropics and temperate zone have led to the development of a variety of unique approaches for assessing and managing inland water environments and their fish stocks over the last few decades. Traditional methods developed for specific fish species have proven to be ineffective, with their use limited to a few big lakes where only one or two species are fished. Most other environments' fish populations were made up of a variety of species and were not susceptible to the population dynamics methodologies utilized in marine fisheries. There has been developed a more thorough ecological method that is better suited to the types of fish communities found in most rivers and lakes. It became evident that focusing exclusively on the biology and behavior of the stock was insufficient to characterize the fishery completely. In recent years, the social, economic and policy aspects of fisheries management have gotten increased attention. These findings, along with the failure of centralized management in general, have encouraged the creation of novel co-management techniques that include the devolution of several previously centralized authorities to local governments or even fishing villages. In the new management approaches, transfers of ownership and control from the government or significant landowners to the fishers themselves are typical.

## 6.3. SOLUTIONS

Several solutions have emerged to assist managers and fishermen in better managing their fisheries by providing action alternatives based on their interests and the priorities of the societies in which they operate. Thus, technologies exist to boost fishing production, such as natural fish stock enhancement or to promote conservation, such as ecosystem restoration (Figure 6.3). The consequences of human activities on the ecosystem within the river or lake basin as well as the distribution of benefits among different members of the community, have been obtained. The implications for fish stock genetic health, particularly in terms of species introduction, fishing and aquaculture costs, have only recently become apparent. Although such tactics are not new, they are part of a new, more integrated approach to environmental and fishery management. In temperate countries, recreational fishing has traditionally been the primary utilization of living aquatic resources. Recreational fishing, on the other hand, is growing more popular in some tropical countries. In Brazil, for example, recreational fishing has exploded in popularity in order to meet the needs of the country's rising urban population. Recreational fishing has a significant social and recreational value and it is frequently a country's primary participatory sport. Many European and North American countries currently follow this catch-and-return policy (Arlinghaus, Mehner, & Cowx, 2002).

**Figure 6.3.** Ecosystem restoration.

*Source: https://www.resilience.org/wp-content/uploads/2019/04/bioregion.jpeg.*

Nonetheless, recreational fishermen and their families may consume the catches and there is evidence that such catches are subsistence in nature in some poorer locations. Recreational fisheries catches can add up to a significant amount of weight that is rarely reported in national catch statistics. Recreational fisheries are also significantly more profitable than commercial fisheries. The value per kg of fish caught by recreational anglers is believed to be ten times more than the same fish sold just for food. All expenses sustained by the fisherman, such as travel, lodging and gear, are included in these costs.

## 6.4. AQUACULTURE

Aquaculture is a method for producing fish and other aquatic organisms in a controlled manner, rather than an end goal. In 1996, inland aquaculture produced more than 13.4 million tons of fish for the global market (Philipose, 1966). This accounts for more than 75% of total inland fisheries yields worldwide. The line between capture and aquaculture is becoming increasingly difficult to draw as fisheries managers and fishermen gain greater control of the fishery through stocking, fertilization, environmental manipulation and other operations. Aquaculture systems can range in size from small to large. Intensive aquaculture is usually dependent on specialized installations, whereas extensive methods can simply be utilized to intensify current catch fisheries. From a sociological aspect, aquaculture and its practitioners have a distinct legal status than simple capture fishermen, because aquaculture implies ownership of both the production facility and the organisms being raised. Competition for water, risks of exotic species introduction and genetic degradation of native fishes due to escapes from aquaculture facilities, damage to wild stocks due to overfishing of young fish for rearing and pollution from pond effluents are just a few of the interactions between aquaculture and capture fisheries. Lakes are chemically and physically stable environments from year to year, but seasonal variations can be varied within a single year.

**Figure 6.4.** Shallow lakes.

*Source: https://3.bp.blogspot.com/-_W7fJAP4__w/WNEiR2yZkJI/AAAAAAAAKnk/xnHGHmDPgqY3jiFQ6kpVG0gcEx7Ex9c0gCLcB/s1600/casey-clear-water-state_s.jpg.*

Shallow lakes (Figure 6.4) are frequently mixed by wind and waves, whereas deep lakes typically stratify in response to temperature differences between the surface and deeper waters in the summer. Fish may be unable to survive if the bottom water or hypolimnion becomes deoxygenated. In any event, nutrients accumulate in this layer during the stratification stage. The stratified condition breaks down as the weather cools in the autumn and an overturn occurs, mixing the water and creating a nutrient pulse. Lakes freeze over in the winter in the cool temperate zones, providing a risk of reduced oxygen beneath the ice. Smaller lakes in the tropics may dry up during the dry season. As a result, lakes at nearly all latitudes exhibit seasonality.

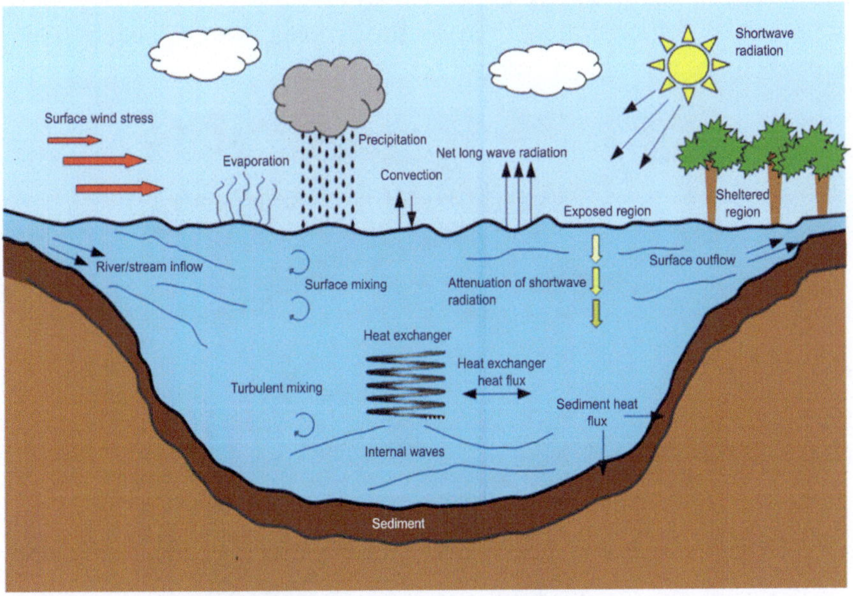


**Figure 6.5.** Hypolimnion.

*Source: https://ars.els-cdn.com/content/image/3-s2.0-B978008100311400008X-f08–01–9780081003114.jpg.*

Trophic richness isn't the sole factor that influences a lake's productivity. Productivity may also be connected to physical traits. Because of the impact of reduced light and temperature at greater depths as well as the deoxygenated hypolimnion (Figure 6.5) which reduces the amount of life that can be sustained by deeper waters, deep lakes are often less productive than shallow lakes in the same region. Shoreline development is important because deep shorelines increase the number of riparian plants and the number of shallow embayments. Flooding arises at different times and in different places depending on the climate zone. When the spring snowmelt releases significant volumes of water from the upstream area, major floods occur in north-flowing rivers in the Arctic. In the temperate zone, spring snowfall on mountain chains at the source can cause rivers to flood, resulting in a predictable pattern of spring floods. Floods (Figure 6.6) can occur at any time of year in reaction to significant rainfall, but they are more likely to occur in the spring. The monsoon or major rainy season, is associated with a regular pattern of unimodal flooding in the tropics. Two floods may occur each year in the equatorial zone as a result of biannual rains. Depending on the type of fish present, rivers can be classified into zones. In temperate rivers, where successions are produced by species replacement, zonation is more easily demonstrated than in the tropics, where successions are established more through accretion.

**Figure 6.6.** Floods.

*Source: https://www.thesundaily.my/binrepository/mk06–01012022-banjir-melaka_2114133_20220114070911.jpg.*

## 6.5. MIGRATORY SPECIES

Some large migratory species migrate upstream from the river's lower reaches to the headwater tributaries to breed and raise their young in the food-rich riffles. In this zone, fish may lay their eggs in riffle pockets or attach their eggs to rocks or submerged vegetation. Some up-river spawners release semi-pelagic eggs that drift downstream with the water until they reach suitable nursery locations on the floodplain. Because of the increasing complexity and range of the habitat, fish assemblages in the downstream regions are substantially more complex. They are divided into three main behavioral associations, which are generally classified using nomenclature devised by traditional peoples along the Mekong River. Whitefish guild is made up of huge, migratory fishes from multiple families that migrate long distances inside river channels between feeding and breeding grounds. The fish may spend their entire lives in the main channel or they may migrate to the floodplains to eat. Whitefish are one-time spawners, spreading a large number of eggs that may remain in place to hatch or may be pelagic or semi-pelagic pushed downstream by the current (Mugimba, Byarugaba, Mutoloki, Evensen, & Munang'andu, 2021). When flooded, blackfish move only locally from floodplain waterbodies to the surrounding plain, then return to the pools during the dry season. They've evolved to stay on the floodplain at all times, with auxiliary respiratory organs that allow them to breathe ambient air or behaviors that allow them to access the well-oxygenated surface coating. They have a variety of complex breeding behaviors that allow them to keep their eggs and newly hatched fry in reasonably well-oxygenated environments.

**Figure 6.7.** Crayfish.

*Source: https://cdn.britannica.com/94/123294–004–8760B736/Crayfish-crustaceans-crabs-lobsters-feature-pinching-claws.jpg.*

Crayfish (Figure 6.7) species fall between the floodplain-dwelling and long-distance migratory guilds. Crayfish make brief migrations from the floodplain, where they reproduce and eat during high water, to the main river channel or the channel's deeper pools during the dry season. These species are less able to survive extremely low oxygen levels, but they do have complex reproductive behaviors that allow them to breed on the floodplain. Some migratory species may be able to survive in the reservoir's shallow upper reaches, which are still accessible to the river. Other species may thrive in the new ecology, modifying their breeding and eating patterns in the process. Reservoir faunas are frequently made up of small parts of river faunas that gain prominence as a result of their new settings. Following impoundment, a regular succession of dominating species occurs, which can last more than a decade before a more or less stable fauna emerges. There is also a burst of productivity during which harvest levels increase as nutrients released by flooding are depleted. To compensate for native species' inability to adapt to the altered conditions, foreign lacustrine species such as tilapias and carps have been introduced into the basin. For management purposes, the size and species characteristics of fish populations are critical. When the populations of species of various sizes are compared in any aquatic body, tiny species are significantly more abundant than large ones. Individual species dynamics in lake fish populations are generally similar to those described for marine species and they conform well to traditional models in that the environment is stable and has no effect on the processes. In river fisheries, the opposite is true, as abiotic environmental factors substantially influence most dynamic processes. In this context, most rivers were most likely flooded frequently in their natural state, but many have now been regulated to the point where they no longer inundate their floodplains. The dynamics of fish populations in such systems, such as canals, are more similar to those of lakes.

## 6.6. REPRODUCTION

Reproduction is usually scheduled to correspond with a seasonal abundance of food supplies, predator protection and abiotic situations that are favorable. These periods are associated with fluctuations in temperature associated with the seasons in temperate latitudes. Fluctuations in rainfall and water level, which constitute the sequence of wet and dry seasons, are more important in tropical latitudes, where seasonal changes in temperature and day length are less evident. As a result, many species in tropical lakes reproduce all year. Species that live in river systems, on the other hand, frequently spawn right before or during flooding, allowing developing progeny to take advantage of food resources. Although some cichlid and other family species are known to spawn throughout the year, albeit at a lower intensity in riverine habitats, spawning during falling floods is rare. Recruitment is normally higher at larger stock sizes, but it often plateaus and begins to decline above a certain stock size. Density independent or dependent processes can be used to determine the link between stock and subsequent recruitment. According to density-independent mechanisms, egg survival and recruitment are unaffected by stock size or the quantity of eggs deposited, resulting in a linear stock–recruitment relationship across all stock sizes. This model must have constraints because resources are finite and no population can grow eternally (Donaldson, 1986).

## 6.7. FLOODING

Flooding intensity is closely connected with dry-season water levels; it's unclear which component of the hydrological regime has the most influence. Despite the fact that the causal mechanisms underlying these empirical associations are unknown, it is obvious that through extreme dry seasons, both fishing and natural mortality are likely to be higher since fish are more vulnerable to capture by both fishermen and other predators. Natural mortalities due to lack of food and severe abiotic conditions will also be higher, resulting in fewer fish surviving to form the spawning population in the following spawning season. In dry years, egg survival and pre-recruit growth are also likely to be less favorable because fish density will be higher, increasing competition for limited resources such as food, predator protection, and ideal spawning substrate.

**Figure 6.8.** Oreochromis esculentus.

*Source: https://ptes.org/wp-content/uploads/2020/07/Oreochromis-esculentus-2-Anton-Wagner.-.jpg.*

During shorter, shallower floods, total nutrient intake is also likely to be reduced, resulting in lower primary and secondary production and fewer feeding opportunities for newly hatched larvae. One of the key flaws in these empirical models is that they assume that spawning stock biomass is unaffected by fishing effort and that changes in fishing mortality are primarily due to the hydrological regime's effect on catchability.

These specializations have resulted in the development of intricate food webs, especially in lakes, where the relative stability of the water body allows for full expression of trophic specializations. When conditions change, this level of specialization might cause complications for the fish. The extinction of *Oreochromis esculentus* (Figure 6.8) from Lake Victoria, for example, has been linked in part to eutrophication, which caused a change from a diatom-based plankton to a phytoplankton composed of blue–green algae. *Oreochromis esculentus* was adapted to a diet of diatoms, whereas an introduced competing *Oreochromis niloticus* was better at digesting blue–green algae (Figure 6.9).

**Figure 6.9.** Blue green algae.

*Source: https://lakes.grace.edu/wp-content/uploads/2021/06/IMG_7940–1-1024x768.jpg.*

Many species tend to be more adaptable in rivers, adjusting dietary sources based on availability. The role of specializations under these circumstances has been debated. In any case, it is assumed that, under normal flood conditions, the amount of food available at high water surpasses the demand of the fish and is not a constraint on their abundance. Researchers evaluated a number of studies on the sources of food, partitioning among specific species and seasonality in floodplain fish feeding habits.

## 6.8. FEEDING BEHAVIOR

Feeding behavior is intimately linked to growth because it is the primary driver of body weight gain. The availability of food, temperature, dissolved oxygen levels and the presence or absence of predators all influence the rate of eating. Temperature has an impact on it as well, since it affects metabolism, stomach clearing rate and swimming performance. The maximum rate of aerobic respiration is limited by dissolved oxygen concentrations, which can have the same effects as temperature. It can be problematic to distinguish between these effects since maximum dissolved oxygen concentrations are temperature dependent. The physiological status of the fish can also influence feeding rates. As fish grow, their feeding rate or consumption may drop or stop entirely. Feeding rates are also likely to be affected by predator numbers (Li et al., 2020). Prey species must be more alert when predator numbers are high, lowering the amount of time available for foraging. Feeding rates are also determined by interspecific and intraspecific competition for food, while interspecific rivalry is thought to be less important among river fishes due to the inherent flexibility in trophic organization among floodplain groups. Intraspecific competition is more likely to be important and it will be substantially influenced by fish density. The occurrence of intraspecific density-dependent growth has critical implications for estimating the dynamics of floodplain fish populations because biomass density varies dramatically both inter-annually and intra-annually.

In all settings except tropical lakes, seasonal changes in these environmental variables impose a seasonality in feeding. The practice of not feeding fish in subarctic and cold temperate water bodies throughout the winter is well documented. In tropical rivers, a comparable process occurs during the dry season, when most species stop feeding almost entirely, a condition known as physiological winter. Much of this effect can be attributed to the need for oxygen in the capture and digestion of food, which may be greater than the oxygen concentration in the water. Other factors, such as overcrowding, could also be at play. Feeding is considered extremely seasonal in the Mekong system, for example, it is connected to fluctuations in food supply and people density. Food resources become abundant during the high-water period, while fish density is at its lowest, allowing for extensive feeding. During the dry season, densities are maximum, food resources in discrete water bodies are limited and so may be depleted and dissolved oxygen levels are often low. Changes in fish condition, assessed as a function of weight in relation to length, can reveal these feeding intensity patterns. The feeding pattern described above is likely to be representative of the majority of species in most rivers. It is not uniform among species or river systems, however, that air-breathing predators such as *Hoplias malabaricus* in Latin America or snakeheads in southern Asia and Africa can continue to feed throughout the dry season due to their ability to consume atmospheric oxygen.

Existing population structures may be unable to fully utilize all food resources available in a body of water. This happens when there aren't enough species to inhabit all habitats and form food webs capable of utilizing all sorts of food, as in the species-poor inland waters east of the Wallace line in Southeast Asia.

It can also happen when current fish populations are unable to adapt to environmental change, such as in reservoirs or rivers with altered flow and morphology. In such instances, the concept of empty niche is commonly used to justify the introduction of new species that can take advantage of the underutilized resource. The concept of a vacant niche is ecologically inaccurate since the niche is a trait of the species, not the environment. However, the concept is incredibly valuable for management, as efficient use of all food resources is a key aspect in increasing yield, if that is the goal.

Because several factors frequently occur at the same time, it is difficult to determine their relative importance. Rapid growth occurs during high-water periods when food resources are abundant, temperature is at its peak and density or competition is at its lowest. In the dry season, when growth is at its slowest, these conditions are reversed. However, among different species and systems, explanations for seasonality of growth are frequently inconsistent or conflicting. Floodplain fish growth sometimes demonstrates significant interannual variation in addition to the intrinsic intra-annual variance indicated.

These fluctuations have been linked to the severity and duration of flooding, as well as the temperature during the dry season. Flooding on a larger scale will likely stimulate more primary and secondary production on the floodplain, as well as introduce more allochthonous food inputs into the system, boosting feeding chances. At the same time, fish numbers are expected to drop, reducing interspecific and intraspecific resource competition, making it impossible to differentiate the effects of the two variables. *Citharinus citharinus* failed to attain sexual maturity during bad flood years in the Senegal river system, achieving less than half of its normal size, resulting in recruitment failure. In Bangladesh, many fish species are exclusively made up of less than one year old fish that attain maturity at the end of their first year (Njoku, Allison, & Jegede, 2009). For these groups, failure to reach sexual maturity due to insufficient growth could have fatal consequences.

The degree to which fish rely on migration to complete their life cycle is also important for fisheries management and biological diversity. Many riverine species appear to be migratory, meaning that if their migration path is disrupted for any cause, they would perish. When the species is sufficiently valuable, such as in Europe and North America's salmonids and sturgeons, every effort is taken to preserve populations through artificial rearing programs. As a result, in many cases, efforts are currently being made to re-establish supplies in regions where they have been eradicated. When rivers are managed, other species appear to be able to adapt to lacustrine conditions. This could have been an adaptation to climatic unpredictability in rivers, allowing some fish to reproduce regardless of whether there was a large flood or not. This behavior appears to have been preserved in species like roach, which were migrants in the original European rivers but have evolved into a more sedentary lifestyle in regulated systems.

Human elements are complicated social and economic groups that extract, process and eat fish, as well as control the fishery's integration into the larger framework of society. The fisherman is surrounded by a complex web of social, financial, ecological and administrative elements that shape his livelihood and must be considered when managing the fishery. The fisherman, as a producer of fish, is the source of a resource flow within the fishery. This flow represents the process by which the fish harvested is transported from the producer to the customer via a series of middlemen who transform and transport the product. Each link in the chain results in an increase in the commodity's price. Many fishermen do not own the fishing spots they visit, nor do they own the equipment or boats they utilize. In these circumstances, the fisherman must either rent the boats and equipment or work as a salaried employee of their owner. Similarly, if the water being fished is privately held, the fisherman may be required to pay a lease or fee to the landowner in order to fish. The owner is sometimes the government, in which case the price is normally in the form of a license. Fishermen may be subject to income taxes. This system usually works against them because fishermen are paid less for their catch than other stakeholders are paid for their inputs. Furthermore, fishermen are prone to accumulate significant debts to boat owners, landowners, gear suppliers and the government, keeping them economically enslaved. One of the goals of effective management is to break this loop and empower fishermen to control their own financial well-being.

Once captured, fish can be kept by the fisherman for personal consumption, sold fresh or preserved to local consumers or bought at a landing site or straight from the fisherman on the water for transport to distant markets. Although some small fisheries may do so, it is uncommon for a fisherman to market catch personally. The intermediary steps between fishing and consuming are usually carried out by individuals who, like fishermen, may work part-time or full-time at these tasks. The majority of fisheries try to remove fish from the water for a specific reason, such as food or the eradication of undesired fish or medication. In certain circumstances, the fish are removed and maintained alive in aquaria for ornamental purposes or to conserve rare species. Natural fisheries rely solely on natural reproduction and feeding to survive. Such fisheries are constrained by the productivity of the water in which they are conducted as well as the target stock's reproductive potential. Biotic factors such as the number of the species, the degree to which it is fished and competition and predation from other species can all influence reproductive potential. Physical aspects of the environment, such as the number of spawning substrates, temperature, the pace of drawdown in reservoirs and the extent of flooding in rivers, may also have an impact.

The majority of such operations are geared at boosting one or a few species that are particularly prized for the fishery's goals. Some interventions, such as fertilization, can have an impact on the entire fish ecosystem. Manipulation of natural reproduction and productivity cycles can have a significant impact on the species composition of a fish population. Stocking may be pursued on multiple levels, but it is originally viewed as a means of compensating for deficits in target species recruitment caused by overfishing or environmental deterioration. The most prevalent type of stocked fishery is one in which juvenile fish are introduced to grow on natural productivity-derived food items. Regular fishing for food or leisure is used to remove the adult fish from the system. There are different solutions available to alleviate various obstacles in fish production. These include the introduction of new species, pond fertilization, waterway engineering and the eradication of harmful species. Natural capture fisheries are increasingly becoming cultural systems as human inputs become more intense. Catch-and-return rules are important to the US and other temperate countries' sports fishing policies, such as the Netherlands and the United Kingdom. Animal rights organizations have questioned if they are inflicting undue distress to the fish. The effects of catch-and-release on fish behavior and populations are poorly known. Furthermore, in stocked systems, returning fish to the water may result in long-term population growth, resulting in stunted fish or even mass mortality.

**Figure 6.10.** Freshwater shrimps.

*Source: https://biofreshblog.files.wordpress.com/2015/04/image-1.jpg.*

**Figure 6.11.** Freshwater mollusks.

*Source: https://upload.wikimedia.org/wikipedia/commons/c/c1/Bithynia_tentaculata.jpg.*

In the lake or riverbank, freshwater shrimps (Figure 6.10) most freshwater mollusks (Figure 6.11) and a variety of fish species can be found. Bottom-set gear is used to catch the species of this type. Between the bed and the surface, pelagic species can be found. These are usually caught with fishing gear that is set in mid-water or on the surface and is not in contact with the bottom. During migration, spawning, some fish gather. Although they can be consumed in huge numbers, their spotty distribution makes them difficult to locate at times. Surface water species can be enticed to cluster in a light-rich location. Freshwater shrimps and many fishes have a more sporadic distribution, while others can be found singly or in tiny, dispersed groups. Seasonal migrations have a huge impact on fisheries. Because of the predictability of the occurrence and the enormous densities of fish sometimes observed in rivers, species movements through river channels and between river channels and nearby floodplains present greater opportunities for capture. Similar chances exist in lakes where fish ascend inflowing rivers to spawn and in the sea were anadromous and catadromous fish travel between saltwater and freshwater environments. Excessive flow can wash away fishing gear or clog it with detritus, making survival tough for anglers. Diverse fish species have different traits that influence their social value. Size, flavor, firmness of meat or some other cultural characteristic all influence their market value. Furthermore, the value of the same species can differ depending on how it is processed and sold. High-value species are typically large and firm-fleshed, with a clean flavor and long shelf life. Smaller species, which decompose quickly in the tropics' high ambient temperatures, are often less valuable. Fish of the lowest value can be reduced into animal feeds like pet food or fishmeal. Because some fishing tactics can only be justified for high-value species, the value of the fish determines the choice of gear. Individual hook and line fishing, for example, can be economically viable for giant salmon with high individual value by individual fisherman, but it is not ideal for capturing fish for bulk processing into meal.

Many types of fishing gear target specific fish sizes. Such gears are often designed with a mesh that allows small fish to pass through, captures intermediate-sized fish and is unable to detain fish longer than a particular length. Mesh selective gears are normally made of netting, however woven traps can also have comparable properties. The lengths of fish that are susceptible to capture in any given mesh size are typically dispersed. Long, thin species will respond differently to nets than short species, while round-bodied species will respond differently to nets than flat species. By altering the balance between the float line and lead line, gill nets can be set slightly above the bottom when fishing for demersal species or anywhere from midwater to the surface for targeting pelagic fish. The nets are frequently set and anchored in place when working inshore in relatively shallow water. They can also be employed as a surface-set drift net in rivers, where they and the boat that operates them float downstream with the water. Several gill nets can be joined end to end in fleets and arranged in complicated, trap-like formations rather than in a straight line. The net's performance can also be boosted by tapping the water on each side of it to scare fish into it, which is forbidden in many fisheries. Commercial netmakers often supply gill nets, which are generally inexpensive and durable. They're also extremely simple to run, and they've given rise to a slew of new fisheries in reservoirs and lakes around the world. Gill nets are highly selective for both larger and smaller fish and their meshes are typically restricted by law. They are one of the most basic and widely used ways for fishing inland waters, and the preferred method in most lakes and reservoirs, due to their ease of building and operation.

They can be manufactured from locally available materials like palm fibers or bamboo as well as commercially available wire mesh or netting. Depending on the target species, traps can be put unbaited or baited. Baits include rotten flesh, dead fish, palm nuts and corn, among others.

Traps are selective for size in the lower size ranges, but will catch any size that passes through the entrance as it grows larger. Traps are typically placed in areas where fish commonly move or assemble, such as river mouths, main river channels, and floodplain access channels.

They can be used in places where other gears aren't appropriate, such as under floating vegetation mats or in relatively shallow water near the shore. Traps can be set up as standalone engines or linked to barriers that direct fish toward the capturing chamber. Traps and barriers can reach from the surface to the bottom, at least at the mouth, and come in different shapes and sizes. Traps, like long lines, can be set from the shore or while wading through shallow rivers, lakes, and swamps. They usually necessitate the use of a fishing craft for transportation and control. The caught fish remain alive in the trap, allowing it to be set and left for several days. Even after they have been abandoned, traps will continue to collect fish. Long sheets of netting are coupled to a head line with floats, a weighted foot line and long warps for hauling in a seine net.

**Figure 6.12.** Beach seines.

*Source: https://marinegeo.github.io/assets/modules/fish-seines/fish_seine_landing_page.jpg.*

Beach seines (Figure 6.12) are set in a semicircle from the shore, usually from a boat. Long warps are then used to drag the net ashore. The operation of large nets necessitates a large team and is costly. They are more commonly held by a single fisherman who employs a number of laborers and are more of a commercial gear than a subsistence one. Alternatively, a fisher may own and run the net. Small nets, which can be operated by wading or as barrier nets across a river, require fewer persons to operate. The movement of the warps across the bottom, which disrupts and guides the fish within the area being worked, contributes to the effectiveness of fishing. Several fish species are strong jumpers and may be able to get away by leaping over the net as it closes. As a result, during the last phases of closing, fishermen hold the float line of the net.

Beach seines are employed in both rivers and lakes. The organizations of fishermen that maintain such locations are usually very protective of them. Beach seines targeting tiny pelagic species can be launched from canoes far out to sea and brought in by long towing ropes linked to the wings (Liu & Cai, 1998). When such fisheries are linked to lights, their effectiveness is boosted. For example, lamps are fastened to rafts and left to burn offshore for several hours in the fishery for *Rastrineobola argentea* in Victoria Lake. They're then carefully hauled in until they're clustered together along the beach. The net is then rigged up around the lamps and dragged in to gather the accumulating fish. Purse seines are used to catch pelagic species that congregate in thick shoals on the surface of huge lakes like Lake Tanganyika's *Limnothrissa* and *Stolothrissa*. They're also commonly utilized to catch large, costly fish like Nile perch. Purse seines are laid from boats in open waters, usually a large ship and a smaller boat. Purse seining is a technique that includes erecting a long net around the target pool. The net's top is normally at the surface, although it might extend several tens of meters downwards. The bottom of the net is gathered together to form a purse that is gradually hauled onboard the vessel, confining the fish to a smaller space from which they can be collected. Purse-seine fisheries are frequently conducted with the assistance of lights mounted on rafts or fishing boats. Purse seines, like trawls, necessitate a significant investment in the form of boats and nets. As a result, they are commercial equipment run by a single owner who hires laborers to operate the boat and nets.

## 6.9. NETS

Scoop nets are made up of sheets of netting that are fastened to a frame and scooped. The netting can be installed in a flat or pyramidal configuration. Scoop nets have been created for a range of fisheries and smaller versions are popular with artisanal and subsistence fishermen since they are relatively inexpensive. Pelagic fish are caught with large pyramidal scoop nets from canoes in Lakes Tanganyika, Malawi and Kariba. Lamps mounted to the boat or on small rafts floating on the water are commonly used to concentrate fish. Scoop baskets and nets, commonly rounded or V-shaped and fixed on a pole, are used in rivers to catch small and immature fish on the bottom or at the surface under masses of floating vegetation. The same net can be employed as a trawl that is held upright towards the bottom and driven ahead. Plunge baskets have open tops and large mouths. The insertion of pockets around the circumference may improve catching efficiency. Cast nets can only be used from the shore or from boats in shallow waters. They can be employed individually or in groups, in which case the fisherman works together to force the fish inwards to a point where the group can cover them with many nets cast at the same time. Cast nets are one of the most frequent inland water fishing gears because of their low cost and mobility. They take some talent to manage, especially when used from a narrow dugout canoe. Artisanal and subsistence fishers are particularly fond of them. Cast nets are selective for smaller fish, and larger, faster-moving fish can get away from the falling net but become entangled in the process.

**Figure 6.13.** Fish parks.

*Source: https://upload.wikimedia.org/wikipedia/commons/thumb/3/32/Hamilton_Fish_Play_Center_pool_jeh.jpg/1200px-Hamilton_Fish_Play_Center_pool_jeh.jpg.*

## 6.10. FISH PARKS

Fish parks (Figure 6.13) are high-yielding capture technologies that first concentrate fish as refuge traps before getting yields comparable to aquaculture. A fish park can operate as a culture system by increasing food availability and offering a breeding place and refuge if it is properly managed and left in the water for prolonged periods of time. Individual fishermen can set up small fish parks with a few meters in diameter, but larger installations with an area of up to a hectare require collaborative effort to set up and maintain. Because they occupy parts of a lake or river on a long-term basis, all fish parks face ownership issues. The social organization of communities that use this type of structure reflects the requirement for huge labor teams to install, maintain and harvest big parks (Bueno, Ostrensky, Canzi, de Matos, & Roubach, 2015). Fishing is, at its core, a social activity. Even a lone fisherman with the most basic gear is bound by a network of customs that governs where and how he can fish. The majority of fishing gear, on the other hand, is more elaborate, implying equally sophisticated relationships. As a result, customary and governmental regulation have a significant impact on fisheries. Some gears are explicitly prohibited because they are too harmful to society, the environment or the economy. Boat-operated gear necessitates an agreement between the boat owner and the fisherman, which is documented in a customary or formal contract. Brush parks and cross-river barriers, for example, necessitate a high level of cooperation that is normally determined by customary usage. In either case, a hierarchy tends to emerge, with the boat or equipment owner on one side and the operatives on the other. There is also a hierarchy among fishers, with those who own and operate larger gear being ranked higher than those who use more basic gear. As a result, each multi-gear fishery is not homogeneous; rather, it is a condominium of multiple fisheries, each with its own set of demands and behaviors. Technical interactions are defined as competition between two fishing gears over a unit stock of one or more fish species. For example, mesh size limitations may disproportionately harm those who employ simpler gear, while barrier gears may capture an excessive proportion of upstream-migrating fish, favoring downstream anglers. Unfortunately, where strong hierarchies exist, individuals higher in the chain of command have a greater ability to influence policies. Many of the decisions made on gear restrictions disproportionately affect the poor fishermen. When drafting fishing legislation and management policies, special care must be taken to preserve the interests of this group.

Questions of ownership and access rights exacerbate these gear-based problems in inland fisheries. The state, collectives such as villages and private individuals possess a large portion of the world's inland water. Large fixed gears, brush parks and fish holes, on the other hand, necessitate some recognition of long-term rights for those who invest in them. Similarly, the time and effort required to clear seine beaches can only be justified if those undertaking the clearing have a fishing license. Similar arguments can be made for systems that benefit from stocking or fertilization. Ownership patterns aren't always consistent and can change seasonally. For example, one group may claim access to a river during floods, while another group may claim the same territory during the dry season. These characteristics, together with current management practices, are tending to limit fishing to a few sets of people who can pursue the twin goals of conservation and intensification. This trend is most visible in nations where recreational interests have clashed with traditional food fisheries. In almost every case, the recreational faction's stronger influence and economic strength has resulted in the food fishery's eventual extinction. In most regions of the world, fishing is a highly seasonal pastime. The temperature regime determines the seasonality of lakes in the Arctic and temperate zones. When the lakes are frozen over, fishing is usually limited. Special gears, such as rod and line operated through ice holes or gill nets coupled to special ratchets that allow them to be stretched between two holes, may be used. Although captures may alter between dry and rainy seasons due to differences in fish behavior, there is less seasonality in tropical lakes. Rivers and canals that have been regulated have a tendency to behave like lakes. Seasonality in the fisheries is much more pronounced in flood rivers and it is influenced by a combination of temperature and hydrological conditions. Freezing is a genuine risk in Arctic and temperate rivers throughout the winter and fisheries must halt at that time. Seasonality, on the other hand, is usually tied to the flood cycle or to the fish's behavioral features. Diadromous fishes, for example, move into and out of rivers at specified times of year, commonly during the spring and autumn/winter flood peaks. In most cultures, a dish isn't considered complete unless it contains meat, poultry, fish or other animal-derived ingredients. For natural or cultural causes, some people consume more fish than others. Geographical factors, such as those seen in small island nations, coastal locations and river and lake riparian zones, can make fish readily available. In some circumstances, the choice of fish is based on personal preference, while in others, it is chosen for its nutritional and health benefits. Others are cultural, based on religious devotion or tradition.

Fish is a decent source of protein as well as minerals like calcium and selenium. Small fish are often eaten whole, and the bones of larger fish are also consumed in some places, contributing to calcium supply, sometimes significantly. Small whole fish contribute significantly more to dietary balance than prepared portions of larger fish. The majority of fatty marine fish, as well as some freshwater species like Nile perch, are high in beneficial unsaturated fatty acids, particularly omega-3 fatty acids. These have been related to a variety of health benefits, including the development of children's neurological systems and brains, the formation of bones, the decrease of blood cholesterol and cardiovascular disease, as well as aiding in the prevention of arthritis and asthma. Dieticians now recommend eating fish at least twice a week, a factor that has raised demand, resulting in higher costs and increasing pressure on fish supplies. Keeping a fish alive is the greatest method to preserve its freshness. Freshwater fish adapt to stress far better than marine fish, and live fish can be found in restaurants. Fish are delivered in specially prepared trucks at metropolitan markets for high-value products. Some animals are prized for their capacity to survive long periods of time after being captured. These are generally air-breathing animals that can live for long periods of time if kept wet. In Europe, eel is sold live in market stalls in this manner, although in the tropics, a variety of snakeheads and catfish can be carried across vast distances in basins and other tiny containers for periods of a week or more.

If the fish cannot be sold alive, its freshness can be preserved by storing it at a low temperature. Icing it immediately after death or even before and storing it in ice for 2 weeks or longer will keep it fresh. Tropical freshwater fish do well on the ice and a 5 kg Nile perch can be stored for up to 4 weeks. While numerous varieties of ice exist, any ice is preferable to none. If you have a choice, little crushed or flake ice is the best. The worst is block ice that has been roughly crushed. Cooling the fish in a mixture of ice and water and then storing it in flake ice is also a good option. Potable water must be used to make all ice. Keeping iced fish in insulated or refrigerated rooms, containers or vehicles is required. Many vessels in freshwater fishing in South America and Southeast Asia are fitted with cold-storage chests and ice is available for suppliers at fish landings or supply stations along the river. Ice supplies are less prevalent in Africa, though they have been constructed in areas with commercial fisheries.

The fish can be frozen if it can't be kept alive or iced. Domestic freezers, plate freezers, and blast freezers can all be used for freezing. Larger fishing vessels and cold storage systems, particularly those linked to formalized marketing networks like supermarket chains, are equipped with freezers. Freezer trucks are used in other fisheries to provide a movable platform for visiting particular landings. Domestic freezers are not suggested for freezing since the process takes a long time and results in big crystals that harm tissues and cause drip losses when thawed. To keep frozen fish from drying out, it should be covered with ice. Alternatively, it can be vacuum-sealed in plastic bags or plastic trays. Temperature and fat content limit storage life, which ranges from 3 months to 1 year.

The construction and maintenance of freezing and ice-producing plants are typically the responsibility of the private sector. The cost of such facilities is significant and in order to recoup the investment, pressure is sometimes placed on the fish resource. Experiments in numerous fisheries, including Lake Victoria, have demonstrated that allowing too many plants to be licensed can lead to increased fishing effort and a long-term decrease in catch. Drying should be done on racks to speed up the process and protect it from dirt, animals and other contaminants. Insects are a major issue in tropical areas. Covering the fish while drying, as well as cleaning and storing in insect-free areas, is prohibitively expensive. As a result, dried fish is frequently extensively infested with insects, resulting in losses due to larvae consuming the fish. Although salting and drying helps to keep insects at bay, it is not a common practice in many locations. Furthermore, significant amounts of salt may be unavailable and the costs may be excessive.

# CHAPTER 7

# Small Scale Fisheries

CONTENTS

## 7.1. INTRODUCTION

Fisheries have grown and diversified at a slow pace over the world, nowhere more so than in developing countries' small-scale industry. Rather than discrete boundaries, we are confronted with a continuum in which scale lines have been bridged, notably through market integration, technology transfer and changing socio-economic contexts. Despite its wide range, the term small-scale remains a relevant descriptor for a set of fisheries characterized by social, structural and institutional factors that affect governability as well as their vulnerable environment. The decentralized character of fishing operations, low levels of capitalization, the household or community focus of company structure and the dynamism of fishing activities in geography, time and technology are all important descriptors. In the face of cumulative pressure on fisheries around the world, the international scientific and development communities are calling for an effective governance to be reshaped to reflect the sector's specificities and vulnerability as well as its potential to generate economic benefits beyond the sector's participants. Reforms must contribute to a more sustainable fisheries industry while also ensuring that the larger benefits of small-scale fisheries or SSF (Figure 7.1) particularly for the poor, are protected and sustained. There is a growing collection of case studies showing the importance of food security (Nozomi Kawarazuka & Béné, 2010).

**Figure 7.1.** Small scale fisheries.

*Source: http://blog.worldfishcenter.org/wp-content/uploads/2016/01/b.jpg.*

## 7.2. FISHERIES

Understanding the livelihoods of fisheries in underdeveloped nations necessitates a different set of measures than those employed in wealthy countries.

While commercial fishing is usually a full-time job in wealthy nations, SSF activities are commonly part of a diversified income or subsistence strategy in poor countries. Those who fish but do not consider themselves to be primarily fishermen may outnumber full-time fishermen by several times in inland systems, such as the Mekong Delta or Bangladesh. Many nations' national demographic and economic census techniques only record primary occupations, which might misrepresent the nature of rural agricultural livelihoods and leave out a major portion of the fishing workforce. National socio-economic household surveys with large sample sizes that allow for deeper insights into livelihood strategies are frequently insufficient to give relevant information on fishing households. The ability of the fishing sector to generate employment in preharvest and postharvest occupations, in particular, is of considerable value. Researchers and the development community have long recognized the significant contribution of women to these sectors. The lack of data on all but primary occupations significantly undervalue the value generated by non-harvest activities. Women are also directly involved in fishing in several nations, albeit it is less prevalent. Part-time involvement is common and it focuses on harvesting methods such as gathering, small-cast or lift-net operation and so on.

**Figure 7.2.** Women in fishing activities.

*Source: https://nereusprogram.org/wp-content/uploads/2018/05/6297384048_ba029ac5aa_o.jpg.*

Women's fishing activities (Figure 7.2) are frequently crucial in the direct supply of protein to families and the risk-averse nature of women's fishing can be beneficial to the household's protein supply stability. Recent research has revealed a shift in the conventional division of labor, particularly in inland water fisheries, where economic necessity and food security take precedence over social norms. These processes, as well as the repercussions of such changes in fishing patterns, must be taken into account while putting in place effective management.

## 7.3. PARTICIPATION

Women's participation in harvest and non-harvest activities, as well as the difficulty of data systems to record these activities, have all contributed to a lack of awareness of women's vital role in fisheries. As a result, policy creation is lacking in thought and decision-making is marginalized. While capturing fish is undoubtedly an essential source of income in underdeveloped countries, the majority of fish workers work on land, transporting, processing, and trading the product. An estimated 82 million postharvest employees raise the overall number of fish workers in developing nations to 118 million. Part-time activities, particularly among family members of fishermen, contribute significantly to these high numbers, as evidenced by project case studies, despite the fact that they are frequently absent from official data. If three million fish employees' dwells in affluent nations, this means that 97% of all fish workers live in developing countries. Upstream employment, i.e., jobs related to input supplies such as boat building, engine and gear production, repairs as well as support services in harbors and at landing locations, are not included in these estimations. Upstream work might contribute another 5–10% to overall employment in Ghana and Senegal. Given that the expected overall production from these fisheries is less than 20%, this may appear counter-intuitive (Ndanga, Quagrainie, & Dennis, 2013). One of the important factors to consider is the scale of inland and maritime capture operations. Small-scale marine activities are defined by accepted national standards to include fishing units that are many times larger than typical inland operations. Aside from differences in fishing techniques, livelihood strategy may also play a role. The idea that specialization is more widespread in coastal fisheries than inland fisheries would lead to higher returns from a larger number of full-time fishers in marine fisheries, albeit this isn't always supported empirically. The importance of inland fisheries for direct food security is one of the most remarkable findings of the research. This is especially striking when compared to large-scale marine fisheries, which produce three times the catch yet only supply around the same amount for domestic consumption. Clearly, the large-scale marine sector contributes significantly to foreign exchange profits and the vast quantities of fish processed into fishmeal and fish oil have an impact on this statistic. Inland catches yielded only a few key export goods.

Despite small-scale fishing's dominance, the large-scale sector in developing countries employs more than three times the number of workers as the large-scale sector in affluent countries. 56% of large-scale marine catches were consumed locally. When the estimated figure for all developing countries was adjusted to incorporate the industrial captures of pelagic fish from Peru and Chile as auxiliary data, the estimated total for all developing countries fell dramatically. Among the case study countries with high rates of non-food use, China converts 18% of small-scale marine fisheries catch to fishmeal, while Thailand and Vietnam divert 20–30% of total fish production to non-food uses. As previously stated, the percentage of fish consumed locally is only one of several indicators that may be used to assess the contribution of fish to food security. In Africa, regional trade has been recognized as having a significant direct impact on food security, and a three-way comparison of domestic, regional, and global trade would be particularly useful. Case studies revealed that, with few exceptions, fisheries data systems in the developing world are primarily concerned with production statistics and that they do so in a fashion that provides only a partial view of the national fishery sector.

## 7.4. PROBLEMS

While pointing out flaws in resourcing, coverage, sampling regimes and amount of disaggregation in particular, this type of data is still useful for assessing components of fisheries that may be sampled efficiently using direct measurement methods. Acknowledging that existing systems often miss a significant amount of fisheries production and that supplementary methods must be used to acquire data on components of the small-scale sector is a vital step that must be taken at a national level. While case studies from the current research have been useful in exposing areas where major improvements in current data collecting might be made, they do not provide a uniform answer to existing challenges. A revamped data system with a high level of disaggregation among fishing fleets and sectors was evaluated in a variety of case studies. In other cases, data from sources other than fishery information systems was used to rescale or supplement production numbers. In the Mekong region of Vietnam, for example, an examination of household fish consumption data generated an estimate of inland fish production that was several times greater than current estimates acquired through fishing surveys. Household fish production statistics from provincial production surveys served a similar function in Thailand.

Inland production in Ghana was estimated using market surveys from Lake Volta. Official inland fisheries production estimates were calculated using direct approaches such as measuring and extrapolating production for specific gear types in each case; it is evident that these methods overlooked significant components of production. The importance of household-level data for assessing small-scale and subsistence output cannot be overstated, especially if collected as part of an agricultural or household census. Most national fisheries organizations responsible for normal fisheries data collection lack the skills and mandate to gather such data; laying out a massive compendium of extra data for collection would be unsustainable.

**Figure 7.3.** Millennium development goals.

*Source: https://www.researchgate.net/profile/Hani-Kim-2/publication/335299485/Figure/fig1/AS:794329194971138@1566394246180/The-UN-Millennium-development-goals-and-sustainable-development-goals-10–11.png.*

However, because of the broader agriculture sector's macroeconomic importance and the desire to establish successful poverty-reduction initiatives, broader household survey systems are frequently well-developed and resourced. The existing value in national household income and expenditure surveys, labor force surveys, agricultural censuses, production surveys was clear and data from these sources was effectively used in a number of project case studies. It was evident that modest tweaks like enhancing data disaggregation or upgrading livelihood descriptors would provide an once-in-a-lifetime opportunity to increase our understanding of fisheries. Existing systems can be greatly improved by ensuring that sample sizes are large enough and sampling frames are specified in such a way that statistically significant analyzes of fishing households and the specific occupational factors relating to fishing livelihoods are possible.

Almost every country has laws and regulations outlining the basic goals of their fishing industry. These rules and legislation will mostly determine how a fishery is managed, but international conventions, global goals such as the MDG or Millennium Development Goals (Figure 7.3) and international and regional cooperation agreements will also have an impact. Traditionally, most fisheries aim to maximize long-term output. This most commonly relates to fish capture, but it can also indicate employment or other societal advantages. The fact that these aims are increasingly being changed to fit values such as democracy, human rights, decentralization, integration, empowerment, accountability and adaptation, among others, is forcing authorities to reconsider their objectives. It's crucial to note that these treatments aren't mutually exclusive and many of them share the basic concepts that are required for more long-term, legitimate and holistic management. Integration and involvement, for example, are commonly incorporated into various viewpoints. However, merely stating that fisheries management should become more holistic, participative or equitable will not suffice in finding practical solutions to SSF management difficulties; we must develop more effective strategies to attain these goals. Understanding the management implications and practical potential of alternative frames of reference is one place to start. We'll look at three related approaches that seem to be best suited to SSFs in developing countries. Explicit integration of ecological objectives in state-based fisheries management is a newer phenomenon that is gaining traction.

## 7.5. MANAGEMENT

Fisheries management is still dominated by property rights issues. Even within this category, there are a variety of viewpoints, ranging from debates over private property rights and the roles played by the market, the state, the judiciary and monitoring and evaluation in defining, distributing and upholding these rights, to determining the conditions conducive to managing communally owned resources including community-based management and co-management. Economic theories of human behavior have had a significant impact on natural resource management. The Gordon–Schaefer model, which assumes a predictable relationship between cost, effort, and benefit has dominated fisheries law, policy, and research. The link between these characteristics is managed in a variety of ways, including limiting total catches, limiting effort and regulating how fish are captured. Stakeholders are still debating how to best allocate and control fisheries property regimes. The multiplicity of SSFs in underdeveloped nations makes such management solutions more challenging to adopt. In the recent decade, the concept of resilience has gained traction in the academic literature on natural resource management. Nonlinear change, unpredictability, thresholds, adaptive management, transformation, institutional learning, vulnerability and adaptation to external drivers are among the concepts grouped under the resilience banner. SSFs, as complex systems, exemplify people's and nature's dynamic and unpredictable interdependencies. SSF residents are subject to the compounding impacts of stressors within fisheries systems, as well as ecological and societal causes beyond their control. Building environmental and human adaptive ability is thus critical to maximizing the conservation, social and economic potential of SSFs in developing countries.

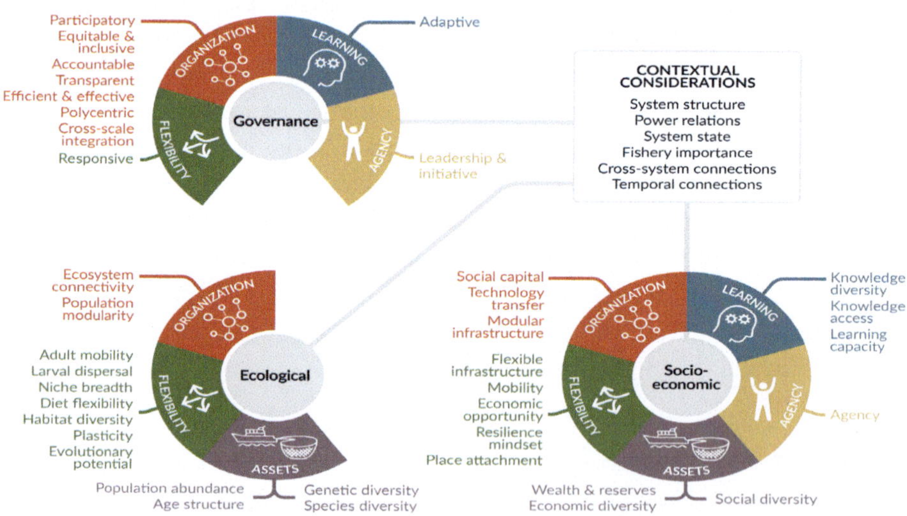


**Figure 7.4.** Resilience techniques.

*Source: https://scx2.b-cdn.net/gfx/news/2021/resilient-fisheries.jpg.*

## 7.6. RESILIENCE TECHNIQUES

Resilience techniques (Figure 7.4) have the ability to significantly improve SSF management when integrated into the EAF's broader legal and policy context. While resilience has become a potent symbol for sustainability, theoretical improvements have yet to be translated into more robust aquatic ecosystems or improved lives for poor fishery-dependent people in developing countries. The major problem currently, especially in developing countries, is to construct bridges between rapid breakthroughs in research and analysis and the policy and organizational restrictions of SSF administration. This challenge is the most critical frontier for SSF research because of poverty and fragility, dynamic non-equilibrium ecosystems and limited capability and data. These difficulties not only pose significant ethical issues, but they also diminish the likelihood of failures being appropriately reported and lessons learned being applied to subsequent ventures. Developing a portfolio of case histories from which to learn requires a decadal time frame. Managers and other decision-makers work inside the framework of government statutes and policies (Bricknell et al., 2021). Most developing nations' fisheries legislation and policy define sustainable exploitation as maximum sustainable yield (MSY) and government ministers and agencies are unlikely to go beyond their legal commitments to test evolving theories. Ecosystem-based management, rights-based methods and resilience perspectives are all mutually compatible. For example, within a broad EAF framework, management for resilience could employ rights-based solutions such as adaptive co-management. Clearly, varied emphases indicate that these approaches or combinations of them, may be more or less successful in other types of fisheries.

The framework used to implement management gives another level of organization beneath the conceptual approach to managing a fishery. It describes the relationships between parts of the research and management challenge and provides a process for solving it. Many features can be found in fisheries implementation frameworks, many of which, while overlapping, represent distinct stages in the process. Scoping, assessment, adaptive management, monitoring and evaluation are all common elements. In a strict sense, an implementation framework is unrelated to management goals, but in practice, both the management approach and the implementation framework contribute concepts and ways of thinking that influence the decisions made. These frameworks, when used together, can guide the diagnosis and management process from the standpoint of any chosen management method, as influenced by international and national policy and regulation relevant to a specific fishery. FAO proposed a broad Fishery Policy and Management Cycle in 1995, with nested levels of activity ranging from international laws and policies through national governance challenges to fishery operational management. A series of feedback loops connect each level to the ones above it, allowing finer-scale and faster-moving processes to be absorbed into the larger and slower levels above.

## 7.7. SMALL-SCALE FISHERIES

Small-scale fisheries (SSF) pose unique management issues since they are diverse in terms of participants, resources and ecosystem services, settings and complicated in their connections to other livelihoods and ecological systems. SSF are similarly subject to external drivers of change, but these aspects are frequently overlooked in traditional fisheries management (Kawarazuka, 2010). To support a broader management emphasis, management innovations that integrate wider system dynamics and improve the ability to better cope with and respond to both external drivers of change and internal sources of uncertainty are required. Small-scale fishing is diverse and difficult to describe. The restrictions and opportunities they face necessitate an emphasis on mastering the fundamentals of management rather than attempting to maximize benefits, as defined historically and narrowly in terms of yield. Absence of research and management capacity in government agencies, political marginalization of fishermen, a lack of quantitative data on fish stock trends and vulnerability to factors outside the fishery are all well-known constraints. The ability of the fishery system to self-organize and adapt, as well as adjust harvest patterns to suit fluctuating resources, arises from the dynamic ecological and social milieu of these fisheries. Diagnostic approaches that strive to contextualize fisheries and explore appropriate entry points are offered instead of blueprint answers for the fisheries management challenge.

Traditional fisheries science and sophisticated management measures that necessitate extensive monitoring and enforcement are unsuitable in developing-country contexts, which are characterized by a lack of data and a limited capacity for intensive management. The amount of money spent on diagnosing processes should be proportional to the value of the fisheries. Many SSF, especially those defined by tiny populations and/or low-value resources, will not be able to afford major investments in assessment and management intervention. However, determining the value of a fishery in order to calculate how much investment is necessary is difficult. SSF's ideals are multifaceted and may not always emerge as predominantly economic value. As a result, these values benefit various stakeholders to varying degrees, impacting who is willing and able to engage in fishery evaluation. In situations when fisheries management is failing and ecosystem services are being harmed, the value of a fishery may not be completely appreciated. At the same time, while the fishery's potential value may be great, the costs of mitigating current threats and reversing existing deterioration, as well as the money available for investment, may limit capacity for diagnosis and management.

SSF diagnosis and management in poor nations, at the very least, necessitates adaptive and collaborative management approaches in which diverse stakeholders are involved in learning and action. Assessment techniques based on many sources of knowledge, as well as reasonably cheap and accessible data and information, can help such management. Rapid and/or participatory assessment procedures, also known as Participatory Rural Appraisal (PRAs), are research tools aimed to reveal the perspectives, knowledge and values of local peoples who are impacted by management choices. Rapid assessments rely on in-person team interaction for data collection and interpretation in order to get a basic grasp of the context from the perspective of the stakeholders. The goal of participatory approaches is to legitimate and improve the relevance of assessments and management. Many approaches to PRA exist, but they are all based on the same core principles: recognizing diverse perspectives, facilitating group learning processes and permitting self-mobilization and context-specific transformation. Participatory methods are particularly adapted to assessing SSF, for example, to answer questions about how and why fisheries stakeholders behave, what system change looks like and what is assumed to be driving it. Participatory techniques are typically used to understand the social and socioeconomic dimensions of resource use systems, but they can also be used to gain local knowledge of fish and ecosystem trends, as well as to support local collection of both ecological and catch data. The costs of participation in gathering such data are less than those incurred in understanding these processes independently of the extraction process.

Once the fishery has been established, the evaluation can concentrate on defining the fishery's biological, social and political setting, as well as the restrictions and opportunities it faces. We must analyze not just the fishery's current characteristics, but also its past and prospective future in order to properly comprehend it, contextualize risk and discover opportunity. Discussing and assessing fishery challenges along time lines in order to reflect past influences on current management is one way to visualize fisheries as dynamic systems. Part of this entails paying close attention to some of the more subtle political and socio-cultural dynamics that continue to shape fisheries management and outcomes, such as property rights, vulnerability and conflict. The larger context of a fishery in underdeveloped countries is frequently considerably different from that in rich countries, where mainstream fisheries science begins. Cultural ideas, traditional customs and even religion might influence the behavior of managers and other stakeholders in such situations. For example, some management strategies are more consistent with Islamic concepts of ownership and use of aquatic resources than others; younger fishers in parts of Africa are hesitant to challenge the authority of elder fishers and thus do not seek leadership positions; and women are frequently excluded from decision-making due to cultural norms. A participatory diagnostic procedure including a diverse group of local stakeholders is more likely to uncover some of these context-specific processes.

Understanding the fishery's history and contemporary context can help to define the dangers and possibilities that it faces. Prioritizing these, in turn, offers a foundation for defining management objectives and performance indices that can be used to track progress in reducing risks and maximizing opportunities. This is a crucial step in the management preparation process. Fishing may be relatively not important to ecosystems in others, particularly inland fisheries, where fishing is part of a diversified livelihood that people enter and leave as needed. Although sedentary invertebrates like sea cucumbers, clams as well as spawning aggregations of long-lived, slow-growing fish are more vulnerable to overfishing than many small pelagic fish or mobile invertebrates, little is known about most SSF to reliable Infrastructure development, such as dams, irrigation projects and roads, looks to represent a significantly greater risk to many river and floodplain fisheries than small-scale, non-capital-intensive fisheries. External drivers can often overwhelm the capacity of fishery stakeholders and management structures to preserve the internal processes necessary for sustainability, renewal, reorganization and adaptation in many of these river and floodplain fisheries. As a result, it's crucial to figure out how much of the fishery's hazards come from within or outside its borders. Clearly, investments in management institutions that focus on fish and fishing dynamics will be ineffective in some cases.

There are several frameworks available that incorporate different aspects of these systems. Fisheries have three basic dimensions: ecological, social and economic. The sustainable livelihoods approach (SLA) looks at fishery-related livelihoods in terms of five capitals. They are natural, physical, human, social and financial and try to figure out how processes, policies, institutions and external shocks influence them. CRiSTAL, an SLA extension, provides a multidimensional framework for assessing hazards connected to climate change as well as opportunities for community adaptive capability. This classification places a larger emphasis on external processes. In the context of this discussion, these categorizations are merely used to group different sorts of difficulties and serve as a reminder to ensure that a wide range of topics is covered in fisheries investigation. These frameworks assist us in considering multiple sources of risk and opportunity. The data needed to clarify the issues that characterize a specific fishery can be gathered quickly from a number of sources utilizing data collecting approaches that are appropriate for SSF in developing countries. These approaches offer an integrated framework for identifying and prioritizing concerns that may be adjusted to each fishery's specific needs. Identifying the spectrum of potential concerns and a technique that gives a qualitative assessment of risks and repercussions are key components of the assessment. The application of ecological risk assessment approaches to social and economic domains, as well as their integration into the management process, is a hot topic that offers numerous advancements in the assessment and management of SSF. Issues must be rated for management purposes once they have been recognized. There are numerous frameworks for developing lists of priorities, management objectives and indicators. Threats and opportunities from different domains of the fisheries are generally ranked and prioritized, resulting in trade-offs and potential conflict. In this stage of the diagnosis, decision-support technologies are provided to aid fishery stakeholders.

Many are extremely analytical and require a large amount of data, therefore they're only useful in the context of SSF. Others, on the other hand, are quite pertinent to our situation. For example, one set of techniques employs risk ranking of broadly defined concerns to determine managerial priorities. Another potential method is the Analytical Hierarchy Process (AHP) (Figure 7.5) which ranks objectives using a series of independent pairwise comparisons of indicators. The goal, regardless of the diagnostic tools utilized, is to identify important threats to the fisheries as well as opportunities to preserve ecosystem services and enable renewal, reorganization and adaptation in response to change. Because the prioritized list of issues should lead management responses, it is critical that it be legitimate and owned by those who bear and manage risk. The people, interactions, and structures that will impact management results are referred to as a fishery's management constituency. The degree to which these are aligned in a given fishery will influence how adaptable, collaborative and legitimate a management system is. The overall management strategy will have an impact on which organizations, institutions and individuals will administer the fisheries. The EAF, for example, results in a wider pool of relevant stakeholders and may require more investment in institution building than traditional target resource-oriented management.

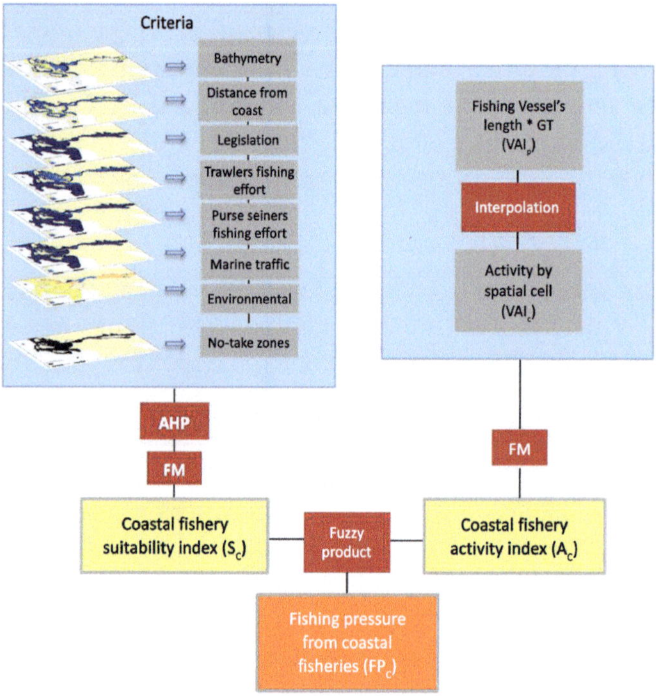


**Figure 7.5.** Analytical hierarchy process.

*Source: https://www.researchgate.net/profile/Irida-Maina-2/publication/282920069/Figure/fig1/AS:320208410562560@1453355042930/Flowchart-of-steps-and-methods-followed-AHP-Analytic-Hierarchy-Process-FM-Fuzzy_W640.jpg.*

## 7.8. APPROACHES

In fact, alternative management approaches will fall short if the management constituency does not sufficiently represent and facilitate the management strategy and the fishery-specific objectives defined under that approach. An important link between diagnostic and efficient adaptive management is the formation of a suitable management constituency for a certain fishery. Stakeholders are people and organizations who are involved in a particular fishing system. Donors, environmental groups, community organizations and tourist operators, for example, who have an interest in a specific fishery but may sit outside the system in the larger governance context, may also be key stakeholders. The term interactions refer to the ties and networks that exist between various stakeholders both inside and outside the SSF. Networks can either reinforce the status quo, impede adaptation or empower and enable people. Institutional, political and economic norms that influence human behavior and can be either restraints or opportunities are referred to as structures. Institutional frameworks will already be in place, power relations will be in play and historical events will have left their imprint.

Proponents of integrated, collaborative and adaptive management as well as organizations like the FAO, push for the inclusion of a diverse but suitable collection of stakeholders. Better problem definition and ownership are expected as well as a more diverse knowledge base for decision-making, increased legitimacy and, as a result, better compliance and commitment to agreed-upon courses of action and conflict resolution. The question is how to find the right group of people to work with. SSF management is at the crossroads of fisheries, environmental protection and development, especially when based on ecosystem-based techniques. Some believe that, as a result, society as a whole has a right to participate in decisions about SSF and its associated ecosystems as global commons. The fact that responsibilities are likely to be scale-dependent and evolve over time makes identifying relevant stakeholders problematic. Nonetheless, after the fishery's physical and administrative boundaries have been established, managers can begin to explore questions to help guide their stakeholder identification process. When the word 'rights' is used in fisheries debates, it conjures up two quite distinct images, depending on one's point of view. First, there are human, social, and economic rights that can be reinforced or negatively impacted by activities performed in the fisheries from the standpoint of persons and communities involved in fishing or otherwise dependent on the fishery. Second, in terms of fisheries management, there are 'fishery rights,' which specify who is allowed to go fishing and who is allowed to participate in the administration of the fishery. This type of right is found in what is known as rights-based fishery management, which focuses on the rights that individuals, communities, businesses, and/or governments have in relation to fishery management. Human and fishing rights have traditionally been considered independently, but there is currently a growing emphasis on integrating the two.

These rights are usually considered in the context of improving management, both by defining who is participating in the fisheries and by involving fishers and others more actively and supportively in the process. Such rights may have an impact on the well-being and security of fishers and fishing communities in a small-scale fishery setting; the effects can be good, given appropriate recognition, design and implementation of rights, but they can also be negative. Human rights and fisheries rights have clear and crucial connections. The previous section focused on the general connections between human rights and fishing rights, but more research is needed to study the various types of fishery rights in greater depth, particularly to assess their implications in SSF. The relevant management authority, whether official or informal, recognizes or assigns rights. In a tribal fishery, for example, the chief may decide who has access to the resource or in another situation, the holders of fishing licenses may be designated by a governmental fisheries body. The rights frequently have a territorial component, in that persons outside the community or region frequently lack access rights and are thus excluded from the fisheries. The situation, on the other hand, is frequently convoluted. To begin with, the anglers in a given location are not always homogeneous.

Furthermore, due to the influence of power elites, the political will to implement these rules and regulations is frequently lacking. While informal and traditional access rights have existed in a variety of fishery jurisdictions for centuries and such rights are being implemented with increasing frequency even where direct government regulations are in place, many SSF are likely to face challenges in making access rights fully effective. SSF managers encounter numerous obstacles. Some issues can be managed, but others are more difficult to forecast or address. Uncertainty is possibly the most difficult of all the issues that managers encounter. There are numerous well-documented sources of uncertainty in natural resource management. When it comes to managing SSF, understanding the margin of error is crucial. At the same time, even if they aren't aware of it, financial managers are frequently operating in a condition of considerable uncertainty. Given the difficulty of ambiguity, fisheries managers may legitimately want to be proactive in reducing both the sources and degrees of inaccuracy in which they work. Even competent managers with all of the necessary technical, human and financial resources to implement and enforce their fisheries management efforts may fail due to uncertainty. More frequently than not, fishery managers lack the necessary technical and financial resources as well as inadequate and/or erroneous data with which to manage fisheries and forecast population trends. To limit the impacts of uncertainty, fisheries management have a powerful instrument at their disposal.

CHAPTER 8

# Dynamics of Aquaculture

CONTENTS

## 8.1. INTRODUCTION

The cultivation of aquatic animals and plants is known as aquaculture. Although growing aquatic food species for human consumption is its main objective, it also has other objectives, like growing fish for ornamental purposes and aquariums. Small rainfed ponds to boost food production for rural families to big commercial farms to supply export items for worldwide markets, aquaculture can be done on a variety of scales. Regardless of its size, aquaculture is an economic activity and the value of its fruits must outweigh their production costs. Because our understanding of the processes and interactions that impact success in aquaculture is less developed than in traditional agriculture, aquaculture tends to be risky and in some cases it is unsustainable. However, aquaculture technology is continuously advancing and more reliable production techniques are becoming available. The majority of aquaculture entails cultivating a target species in conditions that can be monitored and controlled.

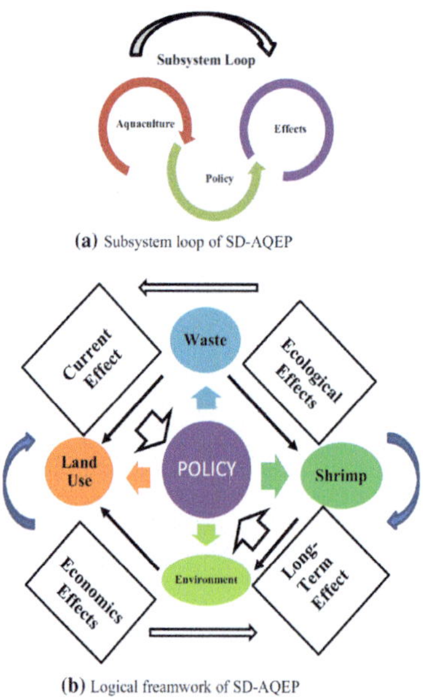


**Figure 8.1.** Dynamics of aquaculture (a and b).

*Source: https://media.springernature.com/lw685/springer-static/image/art%3A10.1007%2Fs10668–020–00594–4/MediaObjects/10668_2020_594_Fig3_HTML.png.*

Mollusks and other sessile organisms can be cultured by providing a substrate for them to adhere to. Fish, crabs and other mobile species, on the other hand, are frequently confined in order to be cultivated. Ponds, raceways, cages and pens are the most typical aquaculture confinements. Ponds are by far the most common method for confining warmwater fish and crustaceans in order to cultivate them (Seawright, Stickney, & Walker, 1998).

## 8.2. CAPTURE FISHERIES

Pond aquaculture (Figure 8.2) can be divided into numerous categories, including species, pond properties and culture intensity. Categorizing aquaculture by species would be cumbersome. For aquaculture species with identical qualities, cultivation processes are usually similar. As a result, it's common to talk about the culture of sport fish, food fish, ornamental fish, crustaceans, mollusks, frogs, other amphibians, alligators and other reptiles as well as vascular aquatic plants, seaweeds and other animals. It's also helpful to categorize the cultures by temperature preference, such as cold water, warmwater and tropical while salinity preference, such as freshwater, brackish water and marine.

**Figure 8.2.** Pond aquaculture.

*Source: https://t3.ftcdn.net/jpg/02/95/91/24/360_F_295912450_LTUjbMdo3mmW0Rhnc1aqx9dZRE7DOZja.jpg.*

Ponds, cages, pens, raceways, tanks and other types of grow-out units can be used to classify culture systems. Watershed ponds, excavated ponds and levee ponds are the three types of ponds defined by construction methods (Belton & Azad, 2012). Watershed ponds are created by damming a natural watercourse when the geography allows for water storage behind the dam. Typically, the dam is built between two hills that confine the watershed. Watershed ponds may receive a combination of overland flow, stream flow and groundwater inflow or they may hold exclusively overland flow. Digging a hole in the ground creates an excavated pond. They can be filled with groundwater inflow near the land surface, overland flow if built in a low-lying area or well water. The water in levee ponds is held in a levee-surrounded region. Because levee ponds receive very little drainage, they must be refilled with water from wells, storage reservoirs, streams or estuaries. Ponds can be classed hydrologically in a variety of ways, such as static ponds with little water exchange or flow-through ponds with frequent water exchange. The most important classification of ponds in terms of pond dynamics is the intensity of management inputs and the amount of production. Nutrient inputs are limited in widespread aquaculture and productivity is minimal. In semi-intensive aquaculture, larger nutrient inputs are provided, resulting in higher productivity. In order to attain very high output, intensive aquaculture requires the highest nutritional inputs.

## 8.3. POND AQUACULTURE

In freshwater pond aquaculture, water exchange is uncommon, but it can be used in very intensive systems. Water exchange is commonly used at all levels of production in brackish water pond aquaculture. Intensive brackish water aquaculture has the highest water exchange rates. As aquaculture grows in popularity around the world, more and more farmers are turning to higher-yielding intensification strategies. For example, more farmers are turning to chemical fertilizers instead of manures and other organic fertilizers in rural subsistence aquaculture. Feeds and mechanical aeration are becoming increasingly important in commercial aquaculture. Production is increasing in both subsistence and commercial aquaculture, but in developing countries, extensive and semi-intensive production techniques are still more common than intensive methods. The term pond dynamics was used by the CRSP to convey the physical, chemical and biological factors that interact in pond systems. In pond aquaculture, ponds are constructed, filled with water and stocked with fish, crustaceans or other aquatic plants and animals. The amount of production in ponds depends heavily upon the quantity of food available to the culture organisms. In systems without supplemental feeding, primary productivity forms the base of the food web that culminates in fish and crustacean biomass. Natural levels of primary productivity are seldom high enough to provide sufficient natural food for high rates of aquaculture production. Some ponds have more nutrients in their bottom soils and waters than other ponds. Thus, there is a varied range in the natural productivity of ponds. In order to enhance natural productivity, manures or chemical fertilizers are applied to stimulate primary productivity.

## 8.4. NUTRIENTS

Nutrient concentrations rise, planktonic blooms develop and dead organic matter falls to pond bottoms as the intensity of aquaculture is raised by the application of manures, fertilizers and feeds. During the day, phytoplankton creates dissolved oxygen through photosynthesis; however, as the biomass of pond biota increases in response to nutrient inputs, respiration increases as well. This causes dissolved oxygen concentrations to fluctuate throughout the course of a 24-hour period, with the highest concentrations in the afternoon and the lowest amounts at dawn in progressive night. Night-time reduction of dissolved oxygen may stress or kill culture organisms if phytoplankton population gets too high. The buildup of organic waste on pond bottoms can lead to anaerobic conditions and the release of microbial byproducts like nitrite and hydrogen sulphide into the water column. Aquatic organisms are poisoned by these toxins. The main nitrogenous waste product produced by aquaculture species is ammonia and as output increases, ammonia concentrations may even become fatal. Low dissolved oxygen concentrations, on the other hand, are frequently the initial limiting factor in pond production. If mechanical aeration is provided to improve the dissolved oxygen level, production can be increased until ammonia toxicity starts to restrict it. The most dependable way to reduce ammonia levels is water exchange. Production levels could be boosted by exchanging water to discharge ammonia from ponds. Because microbial decomposition of stored organic waste on pond bottoms results in deterioration of bottom soil condition and high concentrations of toxic metabolites, production cannot be enhanced indefinitely through aeration and water exchange. The discussion above demonstrates that the growth of aquaculture species in ponds is complicated by a number of ecological factors and interactions. Aquacultural initiatives are significantly more complicated because they take place in a multi-use setting with people living in it. Aquaculture goods must also have a higher value than the cost of production, and this value must be achieved through direct use or marketing. As a result, numerous complicated socio-economic issues are at play in aquaculture, in addition to ecological factors.

Several kinds of tilapia are grown around the world, and this group of fish is becoming increasingly important as a domestic and export species. In 1992, production was estimated to be at 500,000 metric tons. Although some tilapia is produced in water recirculating systems in temperate areas and seasonal ponds in subtropical climates, tropical ponds account for the majority of tilapia production. Tilapias are simple to culture because they are resistant to diseases and poor water quality. They also develop very quickly in congested circumstances, allowing for enormous levels of production. They are suitable for culture in manured or fertilized ponds because they feed on plankton and benthos. They will, however, eat pelleted, manufactured feed and can be reared in high-intensity environments with aeration and water exchange. In developing tropical countries, tilapias are a popular fish to raise. They can be crucial in local markets and the popularity of tilapia among consumers is expanding export opportunities. Tilapias can be cultured at various intensities using techniques that are common to most types of pond aquaculture. As a result, tilapias are an ideal choice for tropical aquaculture research. The findings of tilapia aquaculture research can also be applied to other types of pond fish cultivation.

Pond fish culture encompasses a lot more than just pond maintenance. Pond management, on the other hand, is critical since the majority of the labor and expense goes into growing seed fish into marketable-sized animals in ponds. Mistakes in management can also limit productivity and result in the death of all or part of a pond's fish. Pond management is complicated by a number of variables. They begin with issues such as site selection, design and construction. Pond aquaculture projects may have severe challenges from the start unless a good site is chosen and competent design and construction is applied. The manager must be concerned with climatic and meteorological conditions like temperature, rainfall, sun radiation and wind throughout the grow-out phase. Outside of planning, there is no control over such things. Many other aspects, on the other hand, can be managed to a greater or lesser amount by management. It is possible to choose the culture species and stocking density. Liming, fertilizing and manuring at rates determined by pond soil and water conditions can boost plankton population. When feeding is used, there are frequent options for feed kinds and feeding programs. To treat infections and parasites in aquatic animals, antibiotics and other chemicals are available. Harvest periods can be planned to get the best harvest size, restocking time or market price. Between harvests, pond bottom soils can be exposed to a variety of treatments to improve organic matter decomposition and eliminate pathogens and other undesirable organisms. Mechanical aeration, induced water circulation, water exchange and chemical and biological treatments are all methods for enhancing pond water quality in intensive culture. Determining the proper treatment for a given scenario and the necessary rate at which the treatment should be provided is a major difficulty in pond management. Farmers have created the majority of pond treatments through trial and error. Some of these therapies have only recently been studied to see if they are effective and what the best application rates are. Ponds are extraordinarily complex ecological systems that necessitate careful consideration of the advantages and risks associated with various pond treatments. Unrealistic pond treatment expectations can lead to disappointing yields. For the future development of aquaculture, applied research on pond management and successful extension initiatives for sharing research findings are essential.

Because we are fast exhausting many of the world's resources and causing environmental damage in order to meet the wants of an ever-growing population, sustainability and environmental repercussions are becoming increasingly essential issues in all human efforts. Aquaculture as a business must be sustainable in order to survive. It must, however, be done in a way that does not waste important resources that could be put to better use elsewhere. If done incorrectly, pond aquaculture can have serious social and environmental consequences. Aquaculture operations in wetlands, pond effluent pollution, soil and water pollution from aquaculture solid wastes, social conflicts over resource and space allocations are just a few of the issues that can arise from poorly designed or managed aquaculture projects. Development aid organizations (e.g., USAID, FAO) and several regional organizations have been evaluating the possibilities for aquaculture to address this growing need for fish since the late 1980s. Many of the primary hurdles to greater aquaculture output, according to the literature are issues of policy, institutional development, legislation, property and water rights and human resources rather than research. However, a greater understanding of aquaculture concepts and practices can help in several crucial areas of aquaculture development. Most restrictions, on the other hand, are contextually specified by scale, system and stage of development. Within this framework, research requirements change throughout time and space. They also differ based on the attitude of national or international research funding organizations. A research agenda in Africa directed mostly toward small farmers with limited market access, for example, is fundamentally different from one in Latin America geared toward large-scale, export-oriented commercial growers.

## 8.5. ENVIRONMENTAL IMPLICATIONS

Environmental implications have emerged as a problem, particularly in coastal areas, in tandem with the growth of the shrimp business and industrial expansion in Asia. Water quality and environmental effect, as well as broodstock and seed supply, supplemental feeds and nutrition and natural feeds are consequently general biotechnological research objectives for Asia. As aquaculture becomes more intensive, the focus of research shifts to stress reduction strategies. Fish are kept at higher densities in intensive systems and are frequently fed formulated foods. They are susceptible to disease, which is a rather uncommon occurrence in large-scale pond farming. Indeed, disease has caused very minimal mortality in CRSP ponds during the last decade. Intensification also necessitates a greater investment of time and money to address environmental challenges such as water management and general site features (Tovar, Moreno, Mánuel-Vez, & García-Vargas, 2000). Furthermore, intensive systems are typically aimed toward monocultures and are controlled to satisfy market objectives. To provide a stable product, interest in these systems is focused on genetic engineering and selective breeding. Disease prevention and genetics research is gaining traction in Latin America, as it is in Asia, where aquaculture businesses are fast increasing and systems are becoming more complex. In some areas, however, private sector funding earned from profitable commercial firms may be used for research if the broader public is not deemed to benefit directly.

## 8.6. WATER QUALITY

Water quality is determined by the many chemicals dissolved in the water, as well as the temperature and other physical characteristics of water. Changes in water parameters that promote the production of an aquatic crop are regarded as improvements in water quality, whereas changes that reduce output are termed degradation of water quality in aquaculture systems. This term is significant in aquaculture because the use of water to cultivate aquatic crops at large densities typically leads to chemical properties that are considered water quality reductions by environmental standards. These modifications would not be deemed degradations of water quality for aquaculture purposes unless they reduced the target organism's production, safety or value. Some species will have significantly differing water quality features than others. Characteristics that benefit tilapia productivity may be harmful to rainbow trout. Aquaculture species are frequently chosen for their ability to tolerate poor water quality. As a result, water quality must be considered in the context of the cultured species. An important factor affecting fish output is the water quality of fish ponds. Because the CRSP experimental sites are spread across the globe, the quality of the ambient water varies greatly. Pond management strategies such as fish stocking density, fertilization strategy and supplemental feeding have a significant impact on water quality. Chemical or physical methods such as aeration, fertilizing, liming or water exchange can be used to adjust water quality in an attempt to overcome pond production restrictions. Manipulation of pond water quality is an important management tool in semi-intensive tilapia production and could become a significant constraint in intensive production (Boyd & Pillai, 1985).

**Figure 8.3.** Tilapia.

*Source: https://upload.wikimedia.org/wikipedia/commons/b/b3/Oreochromis-niloticus-Nairobi. JPG.*

Water quality can have an important impact on fish output and can be influenced to some extent by aquaculture operations. Some water qualities, such as dissolved mineral content, pH, alkalinity and hardness are substantially impacted by the watershed's geological and climatic aspects, as well as the source of water and soils. Tilapia (Figure 8.3) thrive in water that is moderately warm and has a high rate of primary production. Water quality may be influenced by other usage of the water and watershed.

Aquaculture techniques have an impact on the quality of water. It is liquid at room temperature other materials that are liquid at room temperature are elemental mercury and various hydrocarbons. Water has the highest viscosity and surface tension of any liquid, with the exception of mercury. The diffusion of gases into and out of the water is severely hampered by this surface tension. Solar radiation does not easily penetrate through water, although it does dim as depth increases. Only 40% of the purest lake water reaches a depth of one meter. The light wavelength has an effect on how well it penetrates water. Penetration is greatest in the middle of the spectrum and decreases as wavelengths go longer (red) or shorter (violet). Infrared radiation heat is mostly absorbed at the surface of a water body and rarely penetrates deeper than 1 m. These qualities cause just surface waters to warm, while wind influences the mixing of heated water into deeper layers. The amount of dissolved and suspended elements in water determines the absolute depth of light penetration. A minimum intensity of light within the photosynthetically active radiation wavelengths (400 to 700 m) is required for photosynthesis, which is typically around 1% of surface levels. The photic zone is defined as an area in the water column where light intensity is at least 1% of the surface level. A Secchi disc or a light meter can be used to assess the depth of light penetration; the depth of the photic zone is roughly twice the Secchi disc depth (SDD) (Figure 8.4). A Secchi disc is a 20 cm diameter weighted disc with black and white quarters painted alternately. It is submerged in water till its image vanishes at the SDD.

SDD varies from a few millimeters to over 40 meters in natural lakes, but it rarely surpasses 1 meter in aquaculture ponds. SDD averaged 50 cm for the majority of a fish grow-out period in Thailand ponds that got minimal fertilizer. However, when considerable amounts of nutrients were supplied, SDD dropped to 25 cm or less due to increasing plankton population. Other CRSP sites yielded different results. SDD averaged 20 to 30 cm in poorly fertilized ponds in Panama, Indonesia and Rwanda, but only 10 to 15 cm in Honduras due to high levels of clay turbidity (Azim et al., 2004). Except in Honduras, SDD remained nearly the same when fertilization rates were raised in these ponds. Organic fertilizer application in Honduras really enhanced SDD by removing clay from the water, as well as primary production rates. A popular management method is to use organic material to eliminate colloidal clay. SDD has long been used as a pond indication of relative phytoplankton abundance. Because plankton richness lowers water clarity and SDD, this association may be strong among ponds within a place or region, but not necessarily across regions. According to CRSP data, this association is weakly present at all locations, but it varies by site due to other factors impacting SDD, such as clay turbidity. In fish ponds, nitrogen availability is critical for primary production, yet numerous nitrogen metabolites, as well as chemical fertilizers like $NH_3$, are harmful to fish. Because phosphorus was once thought to be the limiting nutrient in freshwaters, the significance of nitrogen in freshwater productivity is still being contested.

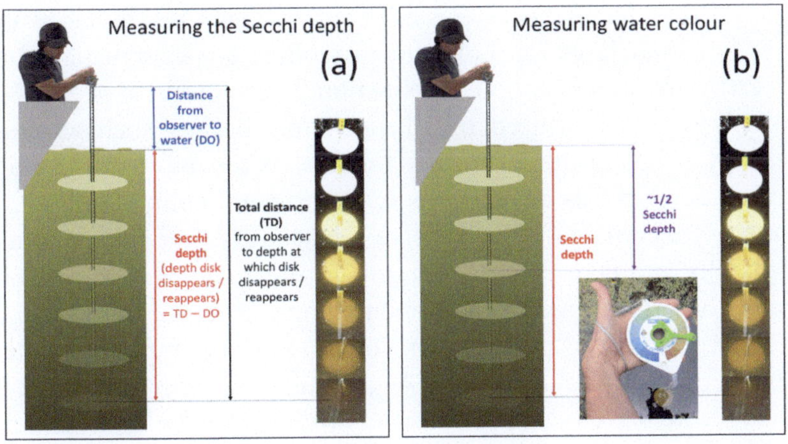


**Figure 8.4.** Secchi disc depth (a and b).

*Source: https://www.researchgate.net/profile/Anas-Abdulaziz/publication/331294915/Figure/fig5/AS:729307328548864@1550891825024/Measuring-Secchi-depth-and-water-color-with-the-mini-Secchi-disk-a-Measuring-the_W640.jpg.*

Some tropical freshwater systems, including aquaculture ponds, appear to be nitrogen-limited. In ponds, inorganic nitrogen is mostly in the form of nitrate, nitrite, ammonia and ammonium. The total inorganic nitrogen (TIN) or dissolved inorganic nitrogen (DIN) is the sum of these products. The existence and quantity of different forms of nitrogen are controlled by pH, oxygen content and organisms that may make or consume certain forms of nitrogen through the nitrogen cycle. Additional nitrogen inputs may occur through phytoplankton and bacteria nitrogen fixation, in addition to nitrogen inputs in these forms from source water, fertilizers and feeding.

When it comes to nitrogen as a nutrient, algal cells usually take it in the form of nitrate, though phytoplankton can also use ammonia. Because of the interconversion of different forms of nitrogen, most pond aquaculture research measures nitrogen as DIN. Input water DIN values range from 0.2 to 0.5 mg/L in Thailand and Rwanda, 0.6 to 1.0 mg/L in Indonesia and 1.5 to 2.0 mg/L in Panama. In all cases, the ammonia-N content of the input waters appears to be larger than the nitrate/nitrite content (Yokoyama & Ishihi, 2010). It has been stressed that nutrients must be delivered in accordance to plankton needs and supplies available from allochthonous sources in order to maintain high levels of primary production. When ponds were treated with chicken dung alone at high inputs, several CRSP tests revealed nitrogen restriction. CRSP fertilization tests have shown that when N and P are both delivered in inputs, large rates of primary production occur, usually at a rate of 4N:1P (by mass) and a total application of 28 kg N/ha/week. Reduced water quality can be caused by nitrogen metabolites such as ammonia produced by fish and other aquatic animals. When inorganic nitrogen or feed are given to ponds at high inputs, detrimental ammonia levels rarely constitute a concern in semi-intensive aquaculture, but they might cause major toxicity problems. Ammonia and nitrite toxicity can be worsened when pH rises. Both unionized ammonia and nitrite are hazardous to fish at low quantities, and as pH rises above 7, unionized ammonia makes up a larger proportion of total ammonia. As a result, a synergy between types of ammonia, nitrite toxicity and pH in ponds with high pH may arise, resulting in toxicity or sublethal stress. This could be especially problematic in lower alkalinity ponds with considerable pH changes during the day. Despite the above-mentioned concerns, lethal or sublethal nitrogen problems have only been recorded in a few CRSP ponds. Under most circumstances, nitrite, and ammonia levels in fertilized ponds stay low. While it's possible that adding urea as a nutrient affects ammonia levels, it was found that employing urea as a nitrogen source in tilapia aquaculture had little negative consequences. When ponds are fed, nitrogen poisoning becomes more frequent.

## 8.7. PHYTOPLANKTON

Phytoplankton is most likely the most important factor controlling pond water quality. Their nitrogen, phosphorus and carbon usage were previously characterized. Photosynthesis generates oxygen, which is essential for the survival of the pond's inhabitants. Phytoplankton respiration during the night has a significant impact on minimum oxygen levels at dawn. Algae can grow rapidly and form a bloom, then die and decompose, resulting in low oxygen levels, which can kill fish or possibly lethal chemicals in the water in some situations. In most fruitful ponds, the relative abundance of phytoplankton confines algal cells to the surface, limiting light penetration. Pond management is crucial for primarily planktivorous fishes, such as Nile tilapia, to increase primary productivity. Phytoplankton biomass in the water at any one time is the result of a dynamic interaction between light, nutrient availability, plankton growth and reproduction and herbivore cropping. Because of their short life cycle and capacity to cause quick changes in pond waters, plankton can display substantial fluctuations in population density from day to day. Algal life cycles can last anywhere from hours to days, allowing for rapid changes in species composition and abundance. Most methods of phytoplankton standing crop in aquaculture do not seek to assess specific species abundance, but rather the overall phytoplankton standing crop. The most generally utilized estimating strategy in CRSP research is to use chlorophyll concentration as a proxy for phytoplankton abundance. Nutrient concentrations and loading rates have a big impact on phytoplankton populations (Smith, 1988).

Tropical aquaculture ponds have a high rate of photosynthesis due to the abundance of nutrients, sunlight and warm temperatures. The generation of oxygen can be used to measure the amount of organic carbon produced by algae. The light-dark bottle approach is the most popular method of measurement, in which pond water is incubated in a clear (light) and an opaque (dark) bottle. In the light bottle, both photosynthesis and respiration take place, whereas in the dark bottle, only respiration takes place. The DO in the incubated light bottle, minus initial DO levels, divided by time equals net primary productivity. The difference between the DO of the initial sample and the sample incubated in the dark bottle is respiration, and the difference between the samples incubated in the light and dark bottles is gross primary production. These bottles may need to be incubated for a long time in unproductive natural waters before meaningful variations can be detected. In ponds, dark bottles may attain zero DO in a matter of hours, whereas light bottles may reach super saturation in a matter of hours. Because of the need for quick incubation, extrapolating the anticipated productivity to an overall daily number is problematic. To compute primary production for the daylight period in CRSP trials, productivity levels measured in the morning are multiplied by a conversion factor based on the proportion of the day's total sunshine that fell during the incubation phase. The light-dark bottle method is challenging to use and understand in aquaculture because of this extrapolation, as well as the quick extinction of light with depth and various other issues. The rate of oxygen production is calculated using variations in DO in pond water over a 24-hour period, corrected for air diffusion. DO measurements are typically performed during different time periods and at various depths and then corrected to produce whole-pond readings each time. Over a 24-hour period, measurements must be taken at least every 6 hours, with more frequent readings near the daily peak time for DO. Another method calculates carbon uptake by chemical equilibria by measuring changes in pH and alkalinity over a 24-hour period. Converting $O_2$ to C using relative molecular weights and respiratory quotients can also be used to assess carbon uptake or productivity. The light-dark bottle approach was utilized in the early CRSP trials, but the free water method was used afterwards. The two techniques have considerable disparities over identical time periods, and there is no meaningful association between these two variables. Both in Thailand and Central America, the free water method consistently yielded higher results than the bright dark bottle method. In the light-dark bottle method, underestimation appears to be a constant bias (Zang et al., 2011).

## 8.8. TROPICAL AQUACULTURE

Tropical aquaculture is frequently carried out in conditions that are thought to be more temperature stable than temperate aquaculture. Tropical ponds, on the other hand, warm substantially during the day and cool dramatically at night, resulting in diel cycles. In addition, most places have rainy and dry seasons, with temperature variations due to cloud cover and other causes. Due to the difficulty of comparing temperature regimes due to daily temperature variations, maximum-minimum thermometers are the most appropriate for this comparison. CRSP data show no predictable seasonal temperature fluctuations or similar daily maximum and minimum water temperatures during wet and dry seasons at all sites. Temperatures in Rwanda were cold enough to prevent tilapia growth and reproduction, whereas conditions in Indonesia and Thailand were ideal for tilapia growth. In shallow tropical ponds, water temperature can show a lot of vertical stratification in addition to seasonal influences. Because of the restricted winds and surficial warming during the day, water stratifies in most CRSP ponds. Stratification was frequently disrupted during the day due to rain or wind. Due to the cooling of the surface waters, stratification failed during the night. Indonesia, Honduras and Panama appear to have stratification patterns similar to Thailand. In Rwanda, the overall temperature is lower, but the temperature difference with depth is even greater, with differentials of 5 to 7 degrees Celsius between the top and bottom layers common. Stratification occurred primarily as a result of warming of the surface waters, with little change in the temperature of the bottom water at all sites.

Automated water quality monitoring allows for many more measurements that are more finely defined in space and time than manual measurements. Pond culture with both smaller or more intense systems uses similar hardware and concepts. Automated monitoring systems can be configured in a variety of ways. There are two broad types of design choices that are dichotomous. One design option is to place sensors in the pond or use sampling pumps to transport pond water to land-based sensors. Pond deployment avoids pumping issues and simply requires the transfer of electrical signals, but it is more difficult to calibrate and maintain. Better sensors, particularly for dissolved oxygen, may make this method more practical and widespread in the future.

The sample delivery system for systems based on land-mounted sensors must be carefully designed to ensure that water quality is not damaged. The second design option is to send signals immediately to a microcomputer rather than to a data recorder for eventual transmission. Pond fertilization has long been used to boost fish production around the world. The use of animal manures and human excreta as main sources of pond inputs for polyculture of the principal Chinese carps is a well-known tradition in China. Inorganic fertilizer has only recently become popular in temperate areas. There is a lot of research on pond fertilization and there are a lot of contradictory and inconsistent results depending on different types of fertilizer, input rates, application methods and frequency. Those opposing opinions could be due to discrepancies in experimental ponds' physical and chemical conditions, as well as disparities in cultured fish species and stocking levels. A lack of proper statistical designs with sufficient replication and common protocols for pond fertilization experiments has also contributed to some of the issues. Pond fertilization and its impact on water quality as well as fish yield were the focus of the shared global experiments. The practical purpose was to give sound pond fertilization procedures and guidance to fish breeders. Work plans for global experiments included simple phosphorus inputs to calibrate sites, comparison of inorganic and organic fertilizers and comparison of various loading rates of organic fertilizers over a period of 2 to 3 years at sites in Honduras, Indonesia, Rwanda, Panama, Philippines and Thailand.

The rate of nutrient uptake by phytoplankton is concentration dependent when nutrient supply is limited. Total phytoplankton production is directly proportional to the initial concentration of the limiting nutrient. However, due to the dynamics of internal and extrinsic elements in pond ecosystems, optimum pond fertilization can be a very difficult task in practice. Individual algae species' differing nutrient requirements, for example, necessitate a well-balanced supply of key nutrients. The most common limiting nutrients for phytoplankton production in natural waters and fish ponds are nitrogen, phosphorus and rare on occasions of carbon. The normal nutritional contents of phytoplankton biomass under optimal development conditions are roughly 45 to 50% carbon, 8 to 10% nitrogen and 1% phosphorus, yielding a typical C:N:P ratio of about 50:10:1. Critical concentration refers to the lowest concentration of a certain nutrient required for optimal growth. Luxury consumption refers to the ability of algae to continue to ingest nutrients over this concentration and deposit the surplus nutrients in their cells without growing. Phosphorus uptake is a well-known example of this process. To maintain this level of output under theoretically ideal conditions, 8 kg N and 0.8 kg P/ha/day are required (Roleda & Hurd, 2019). However, for a variety of reasons, the nutrient inputs required to attain such productivity in ponds are predicted to be lower than the theoretical nutrient allotment in phytoplankton biomass. The algal standing crop in fish ponds is continuously and partially consumed by filter feeding fish, such as Nile tilapia and other herbivorous species in the pond ecosystem, unlike biomass production methods where phytoplankton biomass is completely collected with all nutrients removed. As a result, the amount of nutrients required is only sufficient to compensate for algal growth and grazing and sinking losses. Furthermore, only a small part of the nutrients consumed by fish are absorbed, with the majority being recycled as waste back into ponds.

When it comes to pond fertilization, the N:P ratio is crucial. Although the normal N:P ratio in algal biomass is around 10:1, the fertilizer composition for pond input deviates significantly from this ratio due to the significant differences in natural N and P cycles in ponds. Because of the input from natural nitrogen fixation through biological and atmospheric processes, nitrogen is thought to be less limited in an aquatic environment. Previously, those inputs were assumed to be adequate for low-level pond production and nitrogen fertilizer was considered unnecessary. Phosphorus, on the other hand, is scarce in most natural waterways because it is derived mostly from the weathering of phosphorus rocks, which are not abundant in the lithosphere. Furthermore, dissolved phosphorus availability in natural water is limited by its quick reactivity with cations to form relatively insoluble precipitates that are inaccessible to biota. Pond muds, which significantly adsorb phosphorus, also remove phosphorus from the water column; mud adsorption capacity of P is linearly connected to clay concentration of the sediments. The amount of phosphorus required for pond fertilization is thus controlled by the type of bottom soils and their P saturation parameters, which are frequently linked to the ponds' fertilization.

Excess nitrogen input in the form of ammonia or urea, on the other hand, can have serious repercussions for fish due to ammonia poisoning. Because such fertilizers break down quickly in pond water, ammonia concentrations are expected to rise immediately after application. A single injection of ammonia fertilizer at 5 kg/ha, for example, may cause ammonia concentrations in pond water to approach the stress level of 0.5 mg/L for tilapia. Many factors, including poor weather conditions, severe mud turbidity and nutrient deficits, can induce inefficient nitrogen uptake by phytoplankton. When the proper N:P ratio for a fertilization regime in a given pond is unknown, a phosphorus surplus is preferable to a nitrogen surplus. Before fertilization, if the ammonia concentration in the pond water exceeds 0.5 mg/L, the nitrogen input should be reduced or postponed. As a result, single-grade fertilizers, such as urea or superphosphate, offer more flexibility than mixed-grade fertilizers, especially when just nitrogen or phosphorus is required. Pond fertilization must take into account the financial feasibility of the project. In nutrient-limited ponds, incremental fertilizer application can boost fish output linearly. However, as fertilizer input increases, the biomass gain per unit of fertilizer input will likely decline and fertilization may reach a point of diminishing returns, where the economic return from increased fish weight is less than the cost of additional fertilizer input. Excess fertilization not only wastes resources, but also contributes to poor water quality in ponds and effluents discharged into natural bodies of water. The availability and cost of source materials must be considered while developing a fertilization program for fish growers. Single-grade fertilizers such as urea and TSP may be less expensive than mixed-grade fertilizers, but they are generally harder to come by in locations where the latter is commonly offered for particular land crops. For small-scale farmers, animal and poultry wastes are frequently readily available on-farm fertilizers; artificial fertilizers, which require funds to acquire, may not be a realistic option. The acceptance of animal feces in fish ponds is also influenced by social and cultural factors. In Thailand and many other nations, for example, fish culture utilizing chicken and cattle manures is normal and widely accepted. Human excrement is widely used in China and Vietnam, but it is less accepted in many other nations. Furthermore, for some cultures and religions, swine waste products are forbidden as pond additions.

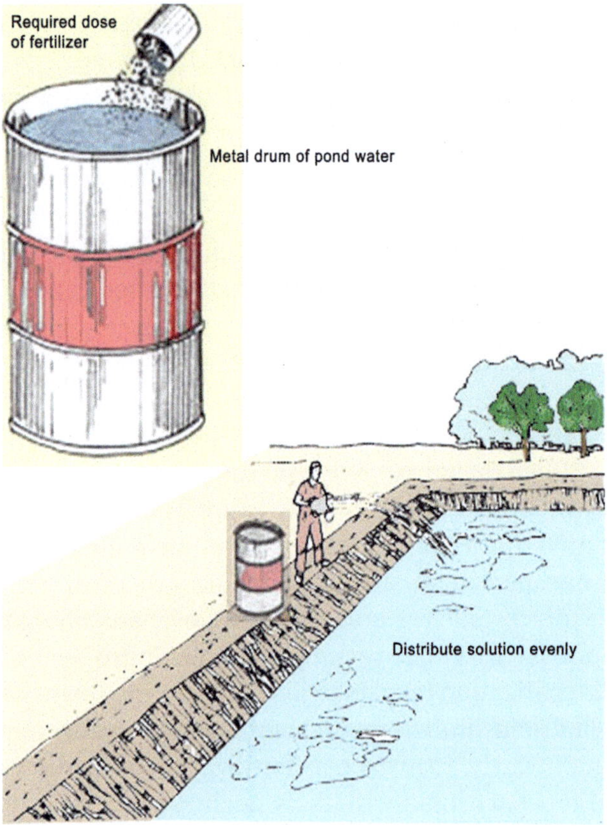


**Figure 8.5.** Chemical fertilizers.

*Source: https://www.fao.org/fishery/docs/CDrom/FAO_Training/FAO_Training/General/x6709e/GR000303. JPG.*

Chemical fertilizers (Figure 8.5) used in fish ponds are similar to those used in agriculture field crops, and their availability and variety are mostly determined by local crop producers' needs. Unlike fertilizers for field crops, which often contain phosphorus, nitrogen and potassium in varying amounts, fertilizers for fish ponds typically do not contain potassium. Potassium fertilization was useless in pond production, the needs for potassium and other micronutrients for pond fertilization have rarely been addressed experimentally. The majority of pond fertilization studies have focused on N and P inputs. For land crops, a variety of chemical fertilizers with varied N and P ratios are utilized. Chemical fertilizer supply and distribution, on the other hand, varies by location, owing to the stage of agricultural growth and the needs of land crops. Traditionally, commercial fertilizers have been calculated using percentages of nitrogen (N), phosphorus ($P_2O_5$), and potassium (K) by weight ($K_2O$). These inorganic fertilizers are granular in nature. Most compound fertilizers contain the key N and P nutrients, nitrate ($NO_3$), ammonium ($NH_4$) and orthophosphate, which dissolve in water as ionic forms of nitrate ($NO_3$), ammonium ($NH_4$), and orthophosphate, respectively. The use of animal manure for fish culture is a natural extension of conventional land-crop farming, which makes use of readily available on-farm resources that are within reach of many small-scale farms in Asia. Small-scale farmers in tropical nations need well-established ways to promote fish culture, which should include fish ponds forming a part of an integrated farming system. Experiments on nutrient budgeting, recycling and energy flow in various components of small-scale farms are needed to make those farms more efficient and sustainable.

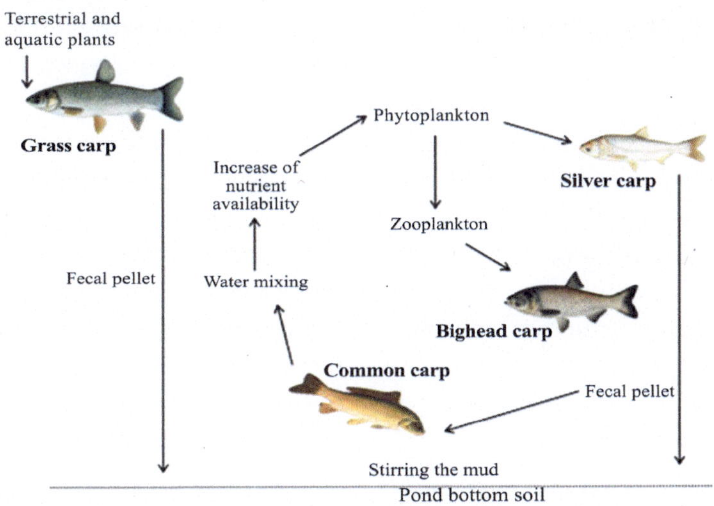


**Figure 8.6.** Polyculture.

*Source: https://www.researchgate.net/profile/Ariel-Zajdband/publication/226829080/Figure/fig1/AS:302276485894145@1449079738533/Polyculture-of-Chinese-carp-in-ponds-and-the-synergies-between-them-Each-species.png.*

It is commonly known that adding animal manure to a pond boosts the production of bacteria, phytoplankton, zooplankton and benthos. Polyculture (Figure 8.6) using a range of fish from diverse feeding niches has been most typically used to maximize fish output with accessible food organisms in ponds. Multispecies polyculture is a harmonious system in which the available fish meals and stocked fish species are balanced. Animal dung as a nutrition source for fish culture has the apparent advantages of being reasonably inexpensive, widely available on-farm and suited for a variety of fish in polyculture. The goal of early fish culturists was to make the pond look like the natural waterways where the fish were located. The need of maintaining cultural conditions that are similar to each aquatic animal's natural environment was realized by early fish culturists. As a result, aquatic species were not farmed in the densities that they are now, although prepared foods were provided in some cases. With regard to the use of clay, adequate sloping of sides, appropriate levee width and techniques to dispose of flood waters, early pond construction procedures were largely the same as those used today. In the early days of aquaculture, the most common pond design was similar to the present raceway. Farmers in the Rhine Valley of France used a slightly different way of production, combining aquaculture and agriculture. Farmers flooded their fields with shallow water and stocked them with fish after a few years of growing crops.

The pond-building processes are based on hundreds to thousands of years old procedures, with just minor changes. Following the design and construction of the pond, the ideas of water budgeting and pond management are discussed as they pertain to the various pond types. Climate considerations, topographical features, water budgets and pond management are all crucial basic elements to understand in order to create a good aquaculture crop. Solar radiation, geothermal energy obtained from the earth's interior as a result of radioactive decay, and gravitational fields contribute to the heating of the atmosphere.

Non-solar energy sources account for less than 0.02% of total atmospheric energy and have only a minimal impact on the earth's atmospheric processes. Solar radiation is absorbed and reflected in identical proportions by land and water, but the reaction to absorption varies due to physical differences between the two surfaces. Specific heat, translucency, evaporation and movement are all differences between water and land. Only a few substances, such as ammonia, have a higher specific heat than water. As a result, if both land and water absorb heat at the same rate, water will heat and cool more slowly. Because water is translucent, solar radiation will penetrate deeply into clear water, heating a significantly larger volume. Because water is more mobile than land, absorbed heat moves both vertically and horizontally within a pond.

## 8.9. TEMPERATURE

Water temperature is generally regulated by seasonal fluctuations in air temperature, which are greatly influenced by seasonal variations in solar radiation. Water density is influenced by temperature, resulting in a natural form of circulation in aquatic systems. Water has a high specific heat, which helps to regulate or withstand variations in water temperature, resulting in rather stable aquatic thermal settings. Solar radiation and, as a result, air temperature varies with latitude and season, resulting in changes in water temperature and pond yields. Other differences between freshwater and saltwater include water density, specific density, heat capacity and vapor pressure, all of which affect their sensitivity to solar radiation and air temperature. Evaporation and radiation are the sources of energy loss from the pond, and they are affected by the season, altitude, latitude, water temperature, wind velocity, pond surface area and relative humidity. The pond soil can also be a source of energy exchange. Shallow ponds, for example, might have a high-water temperature in the summer, resulting in heat loss to the soil. The temperature of a pond can be affected by the flow of water into and out of it; water flow into a pond adds energy, while water flow out of the pond removes it. The placement of the outflow can have a significant impact on pond water temperature, as outlets near the pond's surface tend to remove warmer water, whilst outlets further below may take cold water. The circulation of freshwater ponds varies depending on where they are located.

The fluctuations in water temperature throughout the year cause pond turnover. The pond is frozen over in the winter. Due to heat exchange between the soil and the water just below the ice is around 1 to 2 degrees, while the water closest to the bottom soil is about 4 degrees or slightly higher. The increase in sun radiation in the spring causes the ice to melt. The water temperature closest to the surface will rise to 4 degrees. The surface waters descend to the bottom at 4 degrees because water is densest at that temperature, resulting in the spring turnover. The water in the pond is mixed evenly and has a consistent vertical temperature distribution. The breeze helps with the mixing. The amount of solar radiation increases, warming the water and lowering the density of the surface water. During the summer, winds may be less strong, reducing mixing and resulting in stratification. This stratification results in the establishment of a thermocline, which is characterized by a rapidly dropping temperature with increasing depth if the pond is deep enough. Because the thermocline layer is thicker in early summer than late summer, the temperature gradient is less severe in late summer.

Reduced solar radiation and air temperature, as well as possible increases in mean wind speed, cause surface water temperature and density to drop in the fall. Wind also has a direct effect on water mixing by transferring kinetic energy to surface water, where the resistance to mixing due to the presence of a thermocline is proportional to the magnitude of the thermocline, and as surface water temperature drops, circulation of surface waters becomes deeper until it occurs throughout the pond's depth. This causes the pond to roll over in the fall, resulting in equal vertical temperatures throughout the pond.

## 8.10. RAINFALL

Rainfall is vital in ponds, especially in cases where it is the only way to fill the pond. However, creating a pond where rain is the only source of water is not usually a good idea. Water losses from the pond during droughts result in larger fish concentrations in the pond. This increase in density can result in a variety of water quality issues, as well as the extinction of all fish. Evaporation, seepage, irrigation, weed control and fish management can all cause water to be lost. As a result, most pond culturists choose to supply their pond with water from other, more reliable sources. Rainfall as the primary source of water is usually insufficient to keep pond levels at a safe level. However, if a huge retention pond is later used to replenish ponds, rainfall may be sufficient. The quantity of rainfall in a given place is determined by its geographic location and climate. Rainfall is often concentrated in one season in subtropical areas, but rainfall is spread out over the year in equatorial regions, with one or two seasons of high rainfall. It's crucial to know how much rain falls in a certain area before constructing a pond. Ponds must always be filled to satisfy the needs of the culturist; yet, ponds must not flood, as this would result in the loss of fish. Rainfall can vary greatly across small geographic areas; consequently, a rain gauge should be placed near the pond if the farmer wants a more accurate indicator of the quantity of rainfall it receives (Montes et al., 2022).

# CHAPTER 9

# Developments in Aquaculture

CONTENTS

## 9.1. INTRODUCTION

Many genomic resources, methods and technologies produced through genomics have a wide range of uses in aquaculture research and genetic enhancement of aquaculture species. DNA marker technologies, new sequencing technologies, gene discovery technologies, genome mapping technologies and genome expression analysis technologies are among the most important genome technologies. Even though microsatellites are currently the most common type of marker in aquaculture research and applications, single nucleotide polymorphisms (SNPs) are clearly the markers of choice in the future; the NGS or next generation of sequencing technologies (Figure 9.1) will play a dominant role in sequencing, but may also replace conventional genome expression analysis technologies such as microarrays.

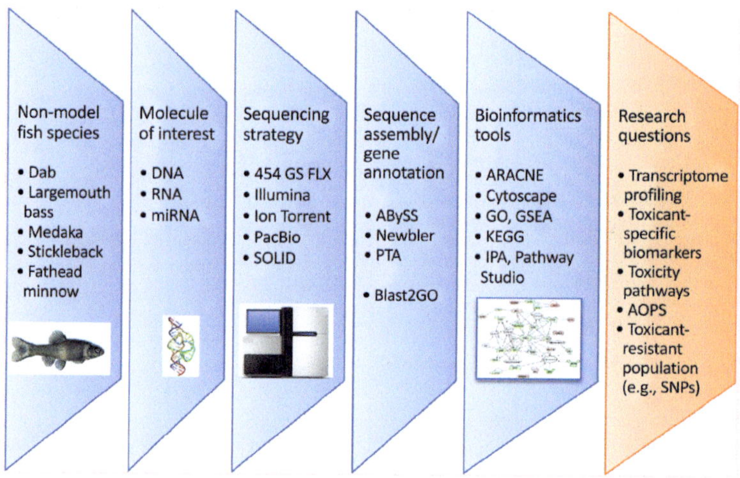


**Figure 9.1.** NGS.

*Source: https://www.researchgate.net/profile/Nancy-Denslow/publication/224856249/Figure/fig1/AS:202500964655105@1425291400208/Flow-chart-of-the-different-steps-involved-in-NGS-based-studies-in-fish-ecotoxicology_W640.jpg.*

Each individual of the same species has a genome that is comparable yet distinct in terms of DNA sequences and encoding capacity, resulting in various transcriptional activities, biological features and performance. The goal of DNA marker technologies is to give a method to detect DNA level variations in genomes between individuals of the same species as well as between related taxa. These assessments have traditionally relied on phenotypic or qualitative markers. Phenotypic markers include morphological variations such as body proportions, size and coloration. It was found that genetic diversity estimations based on phenotypic markers are frequently indirect and inferential through controlled breeding and performance studies. These markers may not accurately reflect genuine genetic differences because they are polygenically inherited and have a poor heritability. Some of these phenotypic markers can only be used to quantify genetic variation once the genetic basis for them is identified. To circumvent the limitations of phenotypic markers, molecular markers such as protein and DNA markers have been developed (O'Reilly & Wright, 1995).

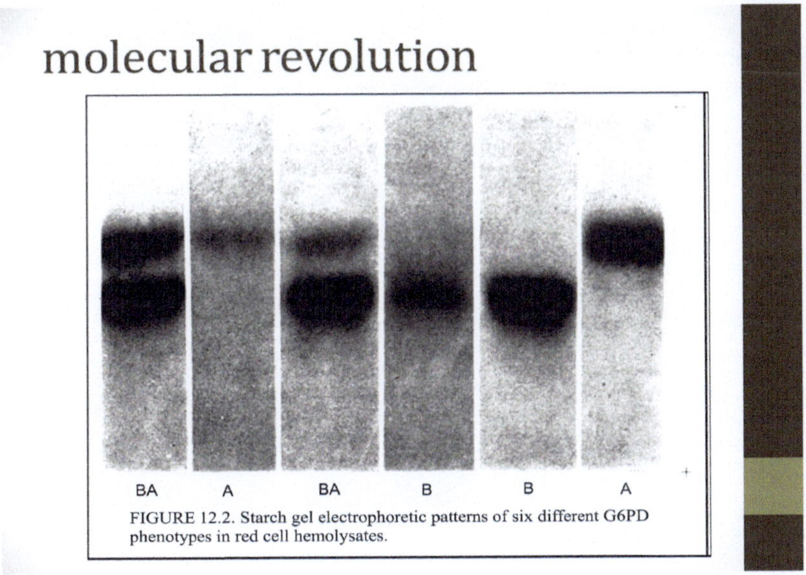


**Figure 9.2.** Allozyme markers.

*Source: https://slideplayer.com/9624746/31/images/slide_1.jpg.*

## 9.2. MARKERS

Allozyme markers (Figure 9.2) were employed to identify broodstocks in fish and other aquaculture species even before DNA markers were discovered. Researchers defined allozymes as allelic variants of the same enzyme encoded at the same locus. Allozyme genetic changes could be the consequence of point mutations, insertions or deletions (indels). In fisheries and aquaculture, allozymes have been used for a variety of purposes, including population analysis, mixed stock analysis and hybrid identification.

However, because of the restricted number of loci, they are becoming a marker type, preventing genome-wide coverage for the analysis of complex features. Furthermore, allozyme electrophoresis may miss a mutation at the DNA level that results in the replacement of a similarly charged amino acid. Another disadvantage is that the muscle, liver, eye, and heart are the most regularly used tissues in allozyme electrophoresis.

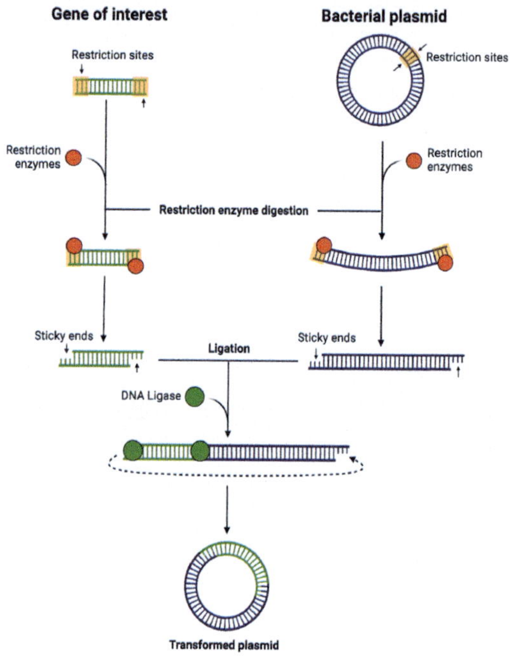


**Figure 9.3.** Restriction enzymes.

*Source: https://prod-illustrations.s3.amazonaws.com/public/templates/5efbb29fe7db1600ac080c71/thumbnail/V2_1597946239127_1007da78–51e3–4d7c-9ef7–026d4fe853e2_public.png.*

The discovery of restriction enzymes (Figure 9.3) in 1973 and the development of DNA hybridization techniques in 1975 laid the basis for the development of the first type of DNA markers, restriction fragment length polymorphism (RFLP). Restriction endonucleases cleave DNA wherever they come across their recognition sequences (Bigger, Murray, & Murray, 1973). As a result of insertions/deletions (indels), base substitutions or rearrangements affecting restriction sites, alterations in the DNA sequence might result in the acquisition, loss or relocation of a restriction site. Individuals, communities and species differ in the amount and size of fragments produced by restriction enzyme digestion of DNA. For RFLP analysis, two methods are commonly utilized. Southern blot hybridization is used in the first, while polymer chain reaction is used in the second (PCR). Southern blot analysis, electrophoresed on an agarose gel, transferred to a solid substrate and detected by hybridization to specific probes, has traditionally been used to separate fragments. Recent research has replaced the time-consuming Southern blot analysis with PCR-based approaches. If flanking sequences for a locus are known, PCR is used to amplify the segment containing the RFLP region. If the polymorphism is caused by a deletion or insertion, the size difference should be seen on gel electrophoresis of the PCR results.

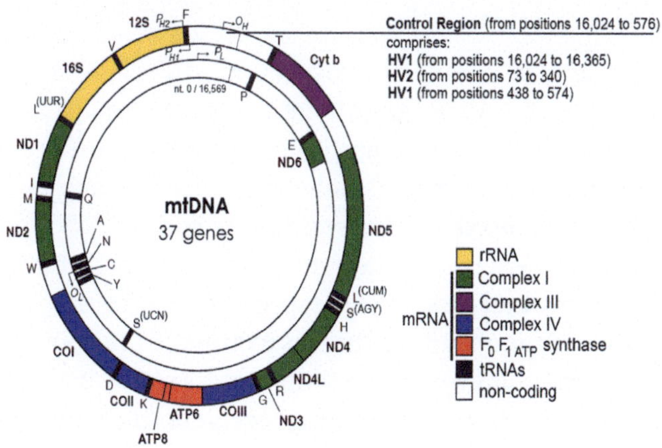


**Figure 9.4.** Mitochondrial genome.

*Source: https://dfzljdn9uc3pi.cloudfront.net/2019/7314/1/fig-1–2x.jpg.*

## 9.3. MITOCHONDRIA

Mitochondrial genome (Figure 9.4) evolution is faster than nuclear genome evolution. Within a species, mtDNA is extremely variable due to its rapid evolution. Because the polymorphism is particularly high in the regulatory region (D-loop region), the D-loop region is particularly relevant in population genetic study. The most common method for analyzing mitochondrial markers is RFLP analysis or direct sequence analysis. mtDNA has been frequently employed as a marker in aquaculture and fisheries due to its high levels of polymorphism and accessibility of mitochondrial DNA sequencing. In most cases, however, mtDNA is inherited from the mother, and this non-Mendelian inheritance severely limits the use of mtDNA in genome study. Furthermore, nuclear genes control the majority of aquaculture-related characteristics. Most farmed finfish species have nuclear genomes in the billions of base pairs range, whereas their mitochondrial genomes are often tens of thousands of times smaller. Clearly, mitochondrial DNA markers will not be particularly valuable for aquaculture genome research and genetic improvement initiatives, notwithstanding their use for identifying aquaculture stocks. However, some recent research has revealed that mitochondrial DNA may have an impact on performance qualities including growth. Alternatives to RAPD that address major issues like low reproducibility were aggressively pursued, and AFLP was the result of these efforts. AFLP is based on utilizing PCR to selectively amplify a subset of genomic restriction fragments.

The adaptor sequence and the restriction site close to it function as primer binding sites for PCR amplification of the restriction fragments. Only restriction fragments will be amplified if the nucleotides bordering the restriction site match the selected nucleotides. Denaturing polyacrylamide gel electrophoresis is used to generate fingerprints from a fraction of amplified fragments. Microsatellites are 1–6 base pair simple sequence repeats (SSRs). Microsatellites are widespread in a wide range of eukaryotic genomes, including all aquaculture species examined thus far. Microsatellites comprise a small percentage of the genome in terms of base pairs included in most vertebrate genomes, depending on genome compactness. More compact genomes, on average, have a lower proportion of repeats, including SSRs, but this is not always the case. The relatively compact genome of the Japanese pufferfish, for example, comprises 1.29% microsatellites, whereas the genome of the closely related *Tetraodon nigroviridis* includes 3.21% microsatellites. Microsatellites were discovered to make up 2.58% of the catfish genome during a genomic sequencing investigation. For every 1.87 kb of DNA in fugu, one microsatellite was discovered. For instance, one microsatellite was discovered in the human genome for every 6 kb of DNA. On average, one microsatellite should exist every 10 kb or fewer of the genomic sequences in most aquaculture fish species.

## 9.4. MICROSATELLITES

Microsatellites (Figure 9.5) with dinucleotide repeats are the most common. For example, dinucleotide repeats account for 67.9% of all microsatellites in channel catfish, trinucleotide repeats for 18.5% and tetranucleotide repeats for 13.5%, eliminating mononucleotide repeats, which aren't nearly as effective for molecular markers.

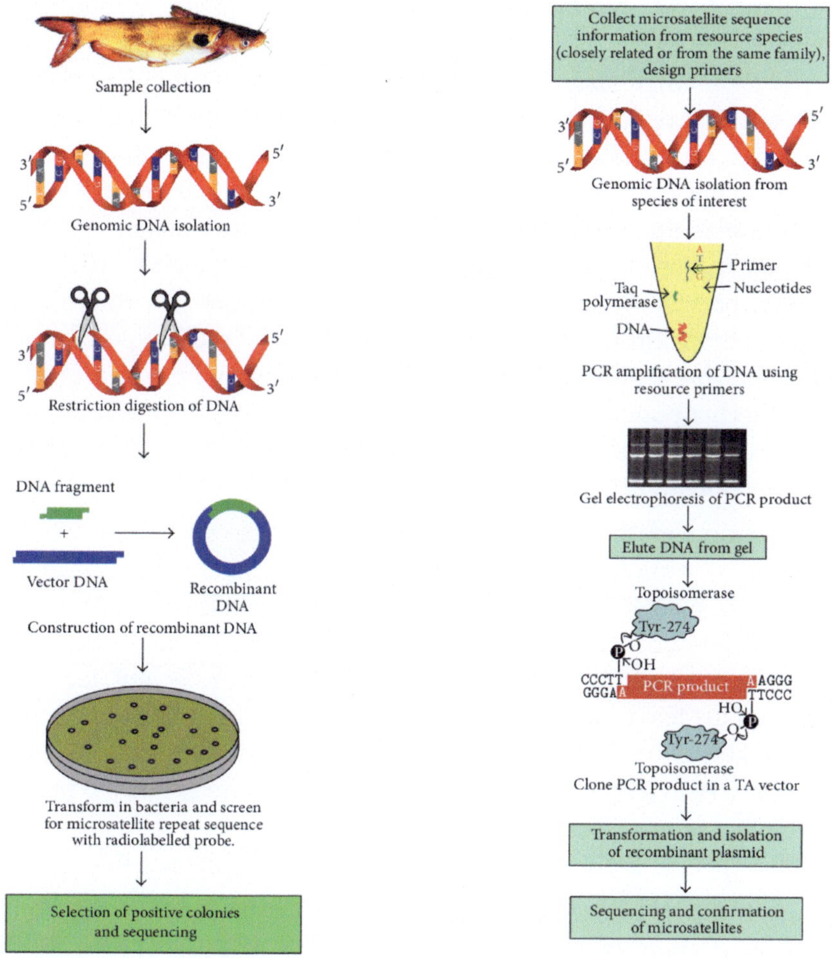


**Figure 9.5.** Microsatellites.

*Source: https://d3i71xaburhd42.cloudfront.net/10583559469bd0bfae57adfcaef aa02660588332/250px/3-Figure2–1.png.*

In general, dinucleotide microsatellites are the most common, followed by trinucleotide or tetranucleotide repeats, while tetranucleotide repeats can be more common than trinucleotide repeats in some conditions. The most prevalent dinucleotide repeat type is (CA)n, followed by (AT)n and finally (CT)n among the dinucleotide repeat types. The (CG)n kind of repeat is uncommon in vertebrate genomes, owing to the fact that vertebrate genomes are frequently A/T-rich. Relatively A/T-rich repeat types are more prevalent than G/C-rich repeat types among trinucleotide and tetranucleotide repeats.

The majority of microsatellite loci are tiny, with only a few hundred repeats. Microsatellite loci are significant for PCR-assisted genotyping because of their small size. Microsatellites with a higher number of repeats are likely to be more polymorphic within a given range, while polymorphism has been seen in microsatellites with as few as five repeats. Small PCR amplicons, usually less than 200 bp, are preferred for the best separation of related alleles that typically differ by as little as one repeat unit. Due to the repeated nature of microsatellites, their flanking sequences can also be fairly basic, preventing the construction of PCR primers for microsatellite loci amplification within a short size limit. Because of their hypermutability, microsatellites are extremely polymorphic, resulting in the accumulation of diverse forms in a population of a given species. Microsatellite polymorphism is based on size differences caused by alleles containing different numbers of repeat units at a given locus.

Multiplexing of their genotyping is possible; however, the extent of multiplexing is limited. Microsatellite genotyping automation is limited, preventing large-scale genome-wide applications. It takes a lot of labor to map thousands of microsatellites to the genome and analyzing tens or hundreds of microsatellites would be a difficult task, if not technically impossible. This leaves just the SNP marker system as a viable option. SNP genotyping is highly automated and thus adaptable to large-scale genome-wide analysis. As a result, it's evident that SNP markers will be the preferred marker in the future. Despite the fact that whole genome sequences for aquaculture species are currently unavailable, it is expected that they will be accessible shortly for important aquaculture species. Furthermore, with the advent of next-generation sequencing methods, it is no longer essential to have the entire genome sequences in order to create a significant number of SNP markers. Genes, habitats, and gene–environment interactions all influence performance and production qualities. Understanding the genes in aquaculture genomes is critical for gaining a complete understanding of performance and production attributes. The predominant method for discovering genes in aquaculture species has been EST sequencing, while other methods like serial gene expression analysis are also available (SAGE). However, the recent adoption of various innovative sequencing systems based on next-generation sequencers has enabled the production of expressed sequence tags by de novo transcriptome sequencing.

## 9.5 CDNA

Single pass sequences of random cDNA clones are known as ESTs. They are incomplete cDNA sequences corresponding to mRNAs derived from cDNA library clones chosen at random. Traditionally, EST analysis has been done by sequencing random cDNA clones from cDNA libraries. This method is effective in the early phases of gene discovery, but it has been shown to be ineffective in the finding of infrequently expressed genes. After reaching a threshold of several thousand ESTs, the pace of gene discovery frequently drops drastically. The most frequently expressed genes would have been sequenced many times before the most infrequently expressed genes were sequenced just once if standard cDNA libraries had been used (Flegel, 2019). Clearly, gene discovery of infrequently expressed genes is inefficient using EST sequencing from non-normalized libraries. Normalization reduces the number of clones that represent abundant transcripts and boosts the efficiency of random sequencing and gene discovery.

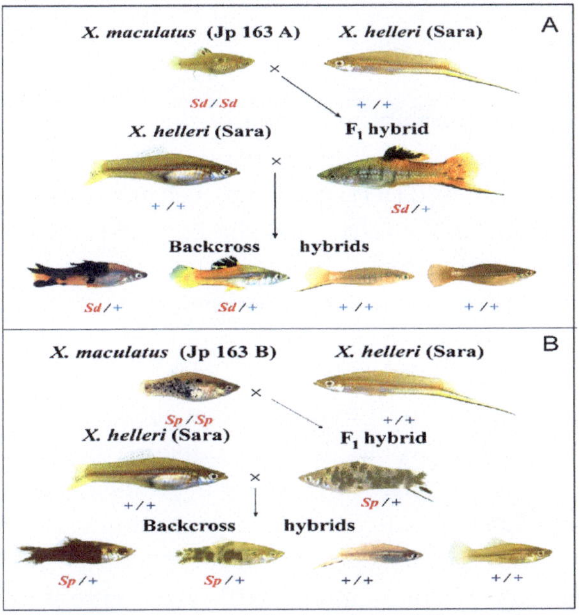


**Figure 9.6.** Linkage mapping.

*Source: https://d3i71xaburhd42.cloudfront.net/26cc7038532a2f7a8392014d89d870586b27817f/3-Figure1–1.png.*

The core of structural genomics is genome mapping and sequencing. While no whole genome draft sequences for aquaculture species are currently available, the genomes of numerous aquaculture species have been mapped. While cytogenetic mapping is extremely beneficial for chromosome identification and physical mapping of genes to chromosomes, it has a low resolution for large-scale genome-wide mapping research. Comparative mapping of aquaculture species has only been done in a few cases.

## 9.6. LINKAGE MAPPING

Linkage mapping (Figure 9.6) can be done with any polymorphic molecular markers that segregate in a Mendelian pattern in the mapping population. Sequence tagged markers, on the other hand, give codominance and transferability across laboratories, as well as potentially beyond species boundaries, allowing for comparative mapping research. Microsatellite markers are the most commonly employed for linkage mapping, while SNP markers may soon overtake them. The examination of marker segregation data within the mapping population can be used to create genetic linkage maps. Various software packages are used to accomplish this. However, it is obvious that the marker densities on aquaculture species' genetic linkage maps are currently insufficient to allow for efficient QTL analysis. The addition of more markers, particularly the SNPs that are now being created, should drastically alter the situation. Somatic cell hybrids were related with radiation hybrid (RH) mapping. The chromosomes of the X-ray radiated cells have been shattered, and the cells are unable to survive on their own. Broken chromosomal fragments from donor cells can be fused into recipient cells after fusion with a recipient cell. A drug-resistant gene can be introduced into the donor cell's genome as a selection marker for the hybrid cells. For growth, fusion cells containing the neo gene would be chosen. Clonal expansion of hybrid cells would result in a panel of RH cell lines, each of which would include a distinct portion of the broken chromosomes, as well as the chromosomal segments encoding the neo gene. Because radiation induces random chromosomal breakdown, the various chromosomal segments found in the panel of radiated cell lines would encompass the whole genome, with some chromosomal segments overlapping. RH mapping is utilized to map the genome using such panels (You, Shan, & Shi, 2020).

## 9.7. PHYSICAL MAPS

Physical maps based on BACs are critical for understanding genome structure and organization, as well as position-based cloning of economically significant genes. A well-defined physical map is frequently used as a foundation for whole genome sequencing. A BAC-based physical map would also enable the use of comparison mapping to harness existing genetic information from map-rich species, speeding up genome research in the species of interest. The creation of large-insert BAC libraries is the initial stage in BAC-based physical mapping. BAC libraries are genomic libraries with a large number of inserts. Large inserts in BAC libraries were created by randomly breaking several copies of the genome with partial restriction enzyme digestion. As a result, the BAC library can be thought of as a collection of genome copies that have been randomly divided into segments that overlap. The presence of the restriction fragments helps align overlapping genome segments since genomic segments from the same genome locations have the same restriction sites. A contig is defined as a set of overlapping genome segments aligned by overlapping restriction fingerprints.

The development of BAC contig-based physical maps for aquaculture species is relatively new. As a result, only physical maps for Atlantic salmon, tilapia and channel catfish have been created, with a physical map for rainbow trout. Physical maps for several of the main aquaculture species are likely to be completed soon. Despite accounting for a large share of global aquaculture production, molluscan shellfish have reaped few benefits from genetic enhancement techniques when compared to other farmed species. Unlike most finfish and shrimp aquaculture, molluscan shellfish aquaculture relies heavily on wild seed regrowth. This is most likely the reason why relatively minor progress has been made in the development and adoption of genetically enhanced broodstock. Overall, hatchery production is less, and as a result, little investment in genetics has been made. Oysters are the molluscan shellfish species that have made the most progress in this regard. The initial genetic enhancements were made in places where wild seed collection is restricted due to environmental conditions or overharvesting and consequent poor recruitment. It's simple to understand how the need for seed sparked the development of hatchery technology, which allowed for controlled reproduction and planned breeding.

Disease-related production restrictions also motivate efforts to develop disease resistance genetically. Polyploidy and, more especially, the formation of triploids, have been the backbones of genetic improvement in marine mollusks thus far. Based on the findings of quantitative genetics studies published since the mid-1990s, one may conclude that selective breeding could significantly boost molluscan shellfish production in the near future. In recent years, several selective breeding projects for oysters have been established and have begun to assist the industry, most notably in the United States, Australia and New Zealand. Selective breeding is a slower method of genetically improving molluscan shellfish than triploidy, which can be accomplished in a single step. It should, however, be viewed as a reasonable long-term complement rather than a replacement, especially because it can target features that aren't affected by triploidization. Genetic gain is cumulative over multiple generations when selective breeding operations are properly planned and maintained, whereas triploidy is a one-step improvement. Triploidy has the benefit of a relatively quick return on investment, but it may also limit or slow future advancement. Because of the increased number and order of genetic interactions, selective breeding of tetraploid broodstocks, which is routinely utilized in oysters to produce triploids by crossing with diploid parents, is known to be slower and more difficult than selective breeding of diploids. As a result, combining selective breeding and triploidy would be a promising but difficult path for genetic improvement.

## 9.8. GENETIC DRIFT

Genetic drift is a well-known process in non-native species transferred to new geographic locations, resulting in a loss of genetic diversity. New species of molluscan shellfish have frequently been produced as a result of such invasions. Following introductions, several occurrences of inbreeding depression have been observed, particularly for scallop species. In 1982, the bay scallop (*Argopecten irradians)* was introduced to China from the United States. Using mitochondrial RFLP, allozyme, and microsatellite markers, the genetic consequence of this problem was revealed. Despite multiple recent introductions of new populations aimed at increasing genetic diversity, inbreeding depression has been established. In 1982, Japan introduced the Japanese scallop (*Patinopecten yessoensis*) to China. In contrast, little or no decline in genetic variety was seen in numerous examples of unintended introduction of non-native oyster species. The Pacific oyster (*Crassostrea gigas*) was introduced from Japan to Australia, New Zealand and France. The Portuguese oyster, which was assumed to have been brought from Taiwan into Portugal and subsequently France, had the similar effect. It was found that feral populations of the Manila clam (*Ruditapes philippinarum*), which were brought from Asia into Europe in the 1970s, have similar levels of genetic variability to local clam species. Hatchery propagation can result in a significant drop in effective population size, owing to the practice of making crosses with a small number of parents. Using allozymes or microsatellite markers, many shellfish broodstocks have been proven to have effective size. Although the use of a small number of extremely fertile species is the most common cause of lower effective size in hatchery-propagated molluscan shellfish, other factors have been discovered.

Inbreeding occurs when the effective population size is reduced, resulting in a limited response to selection in such isolated populations. Inbreeding depression is significantly well documented in molluscan shellfish than a lack of genetic variation. This is due to the comfort with which inbred and outbred progenies may be compared, as well as the small number of multi-generation selective breeding studies. The unmasking of recessive harmful genes causes inbreeding depression i.e., genetic load (Holtsmark, Klemetsdal, Sonesson, & Woolliams, 2008). Oysters have been found to contain extremely high levels of genetic load, as evidenced by significant and frequent distortions of marker segregation ratios. Culling, a frequent method in oyster hatcheries and nurseries, may help to mask inbreeding by selecting genotypes that grow quickly and are more heterozygous. The development of homozygous lines will be slower than in animals with lower genetic load, because the most heterozygous individuals will live and grow better. With the growing interest in the possibility of selective breeding in mollusks, a lot of research has been done on a variety of features in these species. The extent of a trait's genetic base, such as growth or disease resistance, is a direct indicator of its ability to be improved by selection. Such assessments must be made in the context of environmental variation in order to determine the level of genotype-environment interaction and as a result, what value a given group might have in various rearing conditions across different geographical locations. Selective breeding necessitates that phenotypic variance be partially due to genetic components and demonstrating this genetic basis is a requirement for any breeding effort. Estimates of heritability, genetic correlations between phenotypes and genotype-environment interactions are frequently cited as necessary information for establishing a breeding program. However, there are a number of farmed molluscan shellfish species for which these crucial characteristics are still unknown, despite the fact that genetic data may be available.

Aquaculture is a relatively new industrial agricultural activity in most regions of the world, with the exception of Asia's extensive carp culture, Europe and North America's recent expansion of salmonid production. In essence, a genuinely global business has only emerged since the 1970s, focused recently on marine fishes and this sector is looking for ways to construct a dependable and regulated system. Manipulation of photoperiod, water temperature and spawning substrate can all be used to control reproductive function in many fish species. Exogenous hormones are the sole option to create fertilized eggs at an industrial scale in some fish species and they are only employed as a management tool to improve the efficiency of egg production and streamline hatchery operations in others. The reproductive cycle is divided into two phases: growth and maturation, which are influenced by several reproductive hormones produced by the pituitary and gonad glands. The primary motivation for utilizing sterile fish in aquaculture is to reduce the possibility of farmed fish escaping and causing harm to wild populations. Direct consequences of escaped individuals as well as indirect effects of interbreeding between wild and farmed fish and/or the creation of feral populations of farmed fish can be separated into several categories. Only these indirect impacts are addressed through sterilization, which serves as a sort of genetic confinement for any fish that escape from farms. Physical containment is clearly the preferred solution, both for the sake of the farmer and for the sake of wild population protection. As a result, sterilization acts as a safety net in the absence of efficient physical containment. Current aquaculture practices, on the other hand, rarely allow for total confinement of farmed fish populations.

Preventing pre-harvest sexual maturation is another key rationale for utilizing sterile fish in aquaculture. Because of the significant energy involved by farmed fish in gamete synthesis and spawning morphology/behavior, early maturation of farmed fish creates a number of productivity challenges for fish farmers. As muscle energy sources are depleted for the production of gametes, maturing fish decrease flesh quality, especially in females. As a result of their persistent stress, maturing fish have lower immunocompetence and are more prone to disease. Aggressive behavior is common among mature males. For the fish farmer, sterilization can effectively alleviate these issues. However, because maturity age is a heritable feature, they may typically be addressed by selection. Photoperiod manipulation has also been shown to be an effective method for delaying or suppressing maturation in a variety of fish species. Periodic reductions in feeding have produced similar outcomes in salmonids. Another motivation for employing sterile fish is to protect genetic advancements made by selection programs and genetic engineering, both of which are long-term and costly approaches to genetic development. The former is vital to any type of farming, including aquaculture, whereas the latter has yet to be adopted by the aquaculture industry. In either instance, having made such significant long-term expenditures in developing unique genotypes, there is clearly a desire to ensure that producers are unable to develop their own breeding programs based on them. This is a common problem in agriculture and it can be solved by establishing breeding rights through licensing agreements. Sterilization, on the other hand, can act as a safety net if such agreements are not attainable.

The fact that there are alternatives to sterilization for problems like escape, early maturation and the protection of breeding rights explains, at least in part, why sterile fish aren't used more in aquaculture. However, because these options aren't always adequate, there's still a lot of interest in the development and application of effective sterilization techniques. Although there are other methods for sterilizing fish, using female triploid populations on a commercial scale is currently the only acceptable method.

Only germ cells perform meiotic division, a process that reduces a cell's chromosome complement by half such that fertilization occurs in the correct chromosome number. Each diploid premeiotic spermatogonia undergoes two cellular divisions, which are completed by the time of spermiation in males, generating four haploid post-meiotic spermatozoa. Oocytes, on the other hand, delete chromosomes by producing polar bodies, which are lost as the oocytes develop, eventually resulting in the production of a single egg from each diploid oogonium.

Early in ovarian development, the initial meiotic division and polar body production occur, but the second polar body, which has a single set of chromosomes, is preserved until after the oocytes are fertilized. The fertilizing spermatozoon's physical entry of the oocyte causes the oocyte's second meiotic division to complete with the liberation of the second polar body. Thermal treatments are simple to administer and use low-cost equipment, but they are difficult to control, as it is difficult to assure that all eggs in a treatment batch are heated/cooled at the same pace.

The actual treatment temperature is also reliant on the temperature of the pre-treatment incubation, making it challenging to come up with a consistent, standardized therapy in the absence of good incubation temperature control. These issues can be overcome by using hydrostatic pressure therapy, which includes treating eggs in a sealed pressure vessel such that all eggs in a treatment batch receive the same treatment. Furthermore, the optimum pressure treatment is unaffected by the temperature of the incubation chamber. Commercially accessible systems for treating large quantities of eggs with hydrostatic pressure are widely utilized by both commercial and government hatcheries to produce triploids for aquaculture and fisheries management purposes. Hydrostatic pressure treatment for triploidy induction is thought to be quite reliable.

## 9.9. FEEDING

Androgen administration via feeding has been shown to be a cost-efficient and effective technique of permanently sterilizing fish. It is simple to implement on a commercial scale, albeit caution must be exercised to safeguard farm workers and the environment from high steroid doses. The steroid treatments are usually finished one or more years before the fish reach market size, at which point residual steroid levels are no longer detectable. The only reason this technology isn't used in commercial manufacturing is because of consumer concerns about steroid-treated products, and this constraint is unlikely to alter. Shellfish are a popular aquaculture crop. More than a third of global aquaculture production is made up of shellfish species. Oysters, mussels, abalone, shrimp, crawfish, and crabs are all common aquaculture species. In many nations, they help to support important aquaculture businesses. To create superior strains, selective breeding has been used. Biotechnological technologies are increasingly being used to improve the genetics of shellfish stocks.

The alteration of chromosomal sets can be utilized to produce new chromosome structures that could be useful in aquaculture. Diploidy is a phenomenon in which most species, including shellfish, have two sets of chromosomes. We can change the way chromosomes are passed down through the generations, resulting in a wide range of chromosome numbers and compositions. Partial or entire sets of chromosomes may be involved in chromosomal number variations. Aneuploidy refers to a cell or organism that has incomplete sets of chromosomes, whereas euploidy is the condition in which an organism has entire sets of chromosomes. Normal diploidy, haploidy and polyploidy are all examples of euploidy. The paternal origin of chromosomes can also be modified, in addition to chromosome number. By removing chromosomes from one of the parents, uniparental inheritance can be achieved. Gynogenesis and androgenesis are terms for uniparental inheritance based on maternal and paternal chromosomes, respectively. The strict genome organization derived from millions of years of evolution and adaptation is represented by an organism's diploid chromosomal complement. Within species or genera, the number of chromosomes is usually maintained. Most clams and scallops, for example, have a diploid number of 38 chromosomes, but all oysters have 20. Chromosome number changes are significant alterations that can have far-reaching repercussions. Chromosome mutations can be fatal or cause severe phenotype alterations, yet some phenotype modifications can be helpful to aquaculture.

Triploid shellfish, for example, may be sterile and grow more quickly than diploid shellfish. Gynogenesis and androgenesis have the potential to be employed for inbred line production and sex control. Many oyster species have been studied for chromosomal set manipulation because of these and other possible benefits. Female and male gonads in bivalve mollusks like scallops and clams can be different colors. Female gonads in the Chinese scallop, *Chlamys farreri*; Japanese scallop, *Patinopecten yessoensis*; Atlantic Sea scallop, *Placopecten magellanicus* and dwarf surf clam, *Mulinia lateralis* are usually orange to red, while male gonads are milky white. This ability to distinguish sexes is beneficial in chromosome set manipulation because early and accurate separation of sexes can reduce the chance of uncontrolled fertilization or the fertility of most shellfish is extremely high. *Crassostrea gigas*, a two-year-old female Pacific oyster, can generate 20–50 million eggs. Thermal stimulation, UV-treated saltwater, and serotonin can all cause molluscan shellfish to spawn. Female and male shellfish release their gametes into the surrounding water during spawning and fertilization occurs outside the body cavity, with some exceptions. Most crustaceans undergo copulation, in which males deposit spermatophores into females, and fertilization occurs when the females release their eggs. Hermaphrodites include mollusks like Argopecten scallops and Ostrea oysters, which avoid self-fertilization by releasing sperm and eggs at various periods.

Most oyster eggs are in the prophase or metaphase of meiosis I when they are discharged. Fertilization causes meiosis I and II to resume, releasing polar bodies I (PB1) and II (PB2), respectively. Fertilized eggs develop quickly and depending on the temperature, mature into swimming larvae in 10–24 hours. The larvae of most shellfish have a long planktonic larval stage during which they can disperse across great distances. At the end of their larval phase, shellfish larvae commonly undergo some type of metamorphosis. Most shellfish are good candidates for chromosome set manipulation because of their high fertility, external fertilization, and free-living larvae, while other species, such as crabs, freshwater shrimp and oysters, brood their eggs, making chromosome set manipulation more difficult. Disease is still a serious limitation to aquaculture productivity around the world. The risk of new diseases arising and established diseases spreading to different geographical places is a reality as the sector expands and diversifies. The exchange of eggs and fry between fish farms creates ideal conditions for viruses to adapt to their hosts and surroundings. Pathogen control is difficult, requiring a mix of pathogen detection, disease diagnosis, treatment, prevention and general health management. Rapid disease identification and immunization play a crucial role in this, and both have made tremendous progress since the 1980s. Since 2000, the rate has accelerated even faster as clinical and veterinary medicine methods have been rapidly adapted and optimized for use in aquaculture. Novel diagnostic procedures are frequently published, and vaccination has been shown to be effective in lowering antibiotic use, at least in some countries. Alternative approaches for controlling fish infections are also being investigated and implemented.

The continuous losses to the industry caused by pathogens are the main driving force behind improving disease detection and vaccine development. Bacterial diseases are a significant financial burden for the aquaculture industry, and despite the widespread use of antibiotics to control disease outbreaks, there is growing concern about their use due to drug residues in food, the development of antimicrobial drug resistance, and the negative impact on aquatic microbial ecosystems and populations. Because there is a clear need to boost fish production globally through aquaculture, health and welfare must be given top attention if the goals set are to be met. In recent years, some vaccines, particularly multivalent vaccines, have raised concerns about the appearance of adhesions on internal organs, which are suspected to be caused by oil-based adjuvants in the vaccines. Increased production will be aided by the cultivation of new species. Persistent disease concerns in cod have contributed significantly to the stagnation of this new business in Scotland.

## 9.10. PARASITES

Parasite epizootics continue to have significant economic effects in today's aquaculture business, and more effective management strategies are urgently needed. The size of the operations has increased the magnitude of the economic losses, as in the salmon mariculture sector, and the majority of the economic expenses are now attributed to a small number of parasitic infections. For example, the projected cost of sea lice infections to the sector can be as high as 5% of yearly costs in growth alone. Since the 1980s, drugs and chemical treatments have been an increasingly substantial part of aquaculture's production costs. Less expensive medicines are no longer accessible, and the new treatments are specifically suited to aquaculture needs, therefore they are more expensive. As a result of the extensive disease outbreaks, governments are increasingly attempting to control the aquaculture business, albeit this differs from country to country. The main goal is to keep fish from crossing borders in a way that reduces the chance of disease spreading. Since the spread of rainbow trout, common carps, tilapias and eels over the world, the impact of alien parasites on native fauna has been recognized. Without the benefits of evolution, indigenous animals frequently have little defense against foreign parasites and may succumb quickly. Many worldwide pathogenic parasites have been identified, for example, the ciliate *Ichthyophthirius multifiliis*, which originated in the tropics of Asia and now reaches into the sub-arctic areas. Plans for parasitic disease quarantine systems have been developed since the early 1980s and adopted by governments and important bodies, such as the FAO, but implementation at the country or regional level appears to have been sporadic. Thus, spread of fish parasite pathogens across global borders was mostly avoidable, but it was encouraged by misjudgment, which was typically based on a lack of understanding of parasite pathobiology. There hasn't been enough high-level expertise available in vulnerable nations to appreciate the need for, and effectively implement, preventative approaches into the regulatory process. Since the 1970s, expertise has grown fast, and most countries where fish culture is a substantial source of money and/or protein now have some diagnostic and advisory people who can give treatment recommendations for parasite problems.

The possible introduction and spread of parasitic infections through aquaculture activities has garnered the attention of organizations responsible for wild stock protection. As in the case of sea lice, these can be quite contentious. Concentrating research efforts on these issues has benefited the aquaculture business in the long run. For example, there has been significant criticism that mariculturists do not successfully control sea lice infections, which are the source of lice seen on wild salmonids, which is currently a big issue in Canada. This has been a source of discussion and severe political conflict. Some have gone so far as to blame the decline in salmon and sea trout populations on sea lice infection caused by farming activities, to the point where Canada, Norway and Scotland have invested significant research effort in attempting to prove conclusively whether this is the case, and it is a hotly debated topic in these countries. The provision of resources to try to establish new and sustainable systems of control, some of which are now the subject of legislation, is one of the benefits of this dispute. *G. salaris* is a non-virulent parasite of Baltic salmon that has been linked to a huge drop in Atlantic salmon stocks after being introduced into Norwegian rivers. Governments in non-endemic nations have also developed risk analysis and containment plans as a result of their research. For example, the risk of introducing *Gyrodactylus salaris*, mathematical models incorporating such data are increasingly being researched in order to analyze the risk of epidemics through the spread of pathogens across boundaries and to develop strategies for treatment and control. Most crucially, knowing parasite biology and epidemiology has led to parasite control strategies that do not require the use of chemicals. Researchers employed risk analysis to uncover contributing elements in the complex relationships that contribute to the issue. The quantity of Argulus was influenced by low water clarity and high temperature, according to a subsequent examination of the risk variables, which resulted in the infection being managed without the use of chemicals.

For effective administration of chemotherapeutics to the target stage, knowledge of the biology and life cycle of a parasitic pathogen is required. The use of several products that are beneficial against parasitic crustaceans exemplifies the need for targeted therapies. Crustacean parasites have a complicated lifecycle, with molts separating several phases. Organophosphates and insect development regulators are differentially susceptible to these phases, but hydrogen peroxide, pyrethroids and emamectin benzoate have some efficacy against all stages. Excessive, ineffective pesticide use with little economic gain is hazardous to the environment, fish, and the industry's reputation among the public and regulators, as well as contributing to resistance development. Mathematical models, such as the life cycle model developed to anticipate the consequences of targeting therapies against distinct life stages of *L. salmonis*, can help target effective control programs. Researchers used retrospective data on infection levels in Scottish fish farms to advise the industry on a treatment strategy. Molecular epidemiology is being used to study parasite infections, but it isn't widely used yet. However, with the growing availability of genetic data and epidemiological studies, this is a promising topic for future research in fish, as it has been effectively applied to the study of sea lice populations.

In order to better understand host–parasite interactions, a variety of biochemical and immunohistochemical approaches have been used. The interaction between the fish's immunological responses to the parasite and the parasite's immunomodulatory capability/activity is an attractive topic for research that will eventually lead to novel control measures. The situation has greatly improved as a result of the increased number of scientists exploring aquatic pathobiology and innovative approaches. Unfortunately, there are still few taxonomists, although more keys and procedures are available. SEM and TEM are still widely used morphology-based techniques for determining protozoa and other minor histozoic stages of parasites. Specific staining techniques, such as chaetotaxy, and newer technology, such as confocal microscopy and image analysis, support basic morphology. Digitally analyzing forms to extract crucial information has become possible because of advances in image analysis. Following that, the data might be put to a statistical classifier. The automation of the discrimination of *Gyrodactylus salaris* from other salmonid species in the genus has been the most advanced. Recent improvements have demonstrated that these can be semi-automated, preventing the necessity for taxonomists. This is especially useful in situations where pathogen monitoring systems rely on non-specialists and when introducing a single specimen of a species is dangerous, such as with *Gyrodactylus salaris*.

Aquaculturists have a history of using non-chemical approaches only when chemotherapeutics is insufficient or unavailable. The length of the fallow period was calculated using information gleaned from research into sea lice life cycles, as well as economic considerations. Biological control has always been a popular notion, but its implementation has never met expectations. The use of wrasse for sea lice management is perhaps the most successful biological control application for fish parasites. When confined in cages with salmon, members of the wrasse family feed on surfaces such as rocks, but when confined in cages with salmon, they will eliminate lice from the surface of the fish. In the 1990s, this strategy was popular in Norway and Scotland, but it has since dropped due to a variety of drawbacks with wrasse that are costly to overcome. These are primarily linked to the lack of domestication of the animal. Although there is room for improvement, little progress has been made in the use of wrasse since the mid-1990s. Wrasse is best used in conjunction with a variety of different strategies aimed at assisting young and newly stocked fish during their first few months at sea, and it works well with a well-coordinated integrated pest management (IPM) strategy for sea lice. Pathogenic microorganisms, bacteria, fungus, viruses, protozoa, nematodes and other pathogenic microorganisms are used to make biopesticides that are unique to a target pest. For *L. salmonis*, preliminary research of possible biopesticides have previously been conducted. Natural pathogens have been detected and viral and bacterial pathogen research have progressed. Apart from wrasse, biological management has not been adopted by the business, despite the fact that it is widely favored by environmentalists and organic culture agencies, such as the Soil Association in the UK. However, with more investigation, the development of a biopesticide is a distinct possibility. Although no such system has yet been built for an aquatic production system, the challenges are comparable to those faced by terrestrial systems, including specificity, large-scale production, stabilization, storage, product application registration, and so on. Semiochemicals are smells that are distinctive to a species and help locate the host.

CHAPTER 10

# Challenges in Fish Farming

CONTENTS

## 10.1. INTRODUCTION

The existing food supply system, which is built on intensive agriculture, has reduced the number of plant and animal species in our diet while also contributing to natural resource depletion. Animal protein production takes up around 37% of all agricultural land on the planet. Vegetable proteins also require less water, land, nitrogen and fossil fuel to produce than animal-derived proteins. As a result, sustainable techniques have been recommended to reduce environmental impact. Indigenous pulses and root crops, grains, insects, lower organisms, marine algae or by-products from various industries are some of the very nutritious protein sources that are still underutilized. Consumers are increasingly interested in safe, nutritious and healthful food products, in addition to environmental issues. Furthermore, an excessive consumption of proteins from terrestrial animals, which is linked to saturated fatty acids and cholesterol, is thought to be a risk factor for the development of some chronic diseases. On the one hand, aquatic products are thought to have a variety of health benefits, and as a result, the production of various edible marine species for human consumption through aquaculture techniques has augmented in recent years, resulting in large volumes of by-products from fish processing. Healthy bioactive chemicals from natural sources, notably those of marine origin, have, on the other hand, risen in prominence in recent years. The marine environment constitutes a large number of plants, animals and microorganisms that have adapted to such a wide range of environmental conditions that the substances they produce for survival exhibit a diverse range of biological activities (Føre et al., 2018).

## 10.2. FISH PRODUCTION

Around 171 million tons of fish are produced worldwide, with aquaculture accounting for 80 million tons. In 2016, a wide range of species were cultivated in aquaculture, including common carp, Nile tilapia, bighead carp, Catla, Atlantic salmon and rainbow trout. Fish accounted for around 17% of all animal protein ingested by the world's population in 2015. Furthermore, peptides, amino acids, omega-3 long-chain polyunsaturated fatty acids (PUFAs), vitamins and minerals are all found in seafood. Depending on the species, age, gender, health, and harvesting season, the fish has 15–30% proteins, 0–25% fat, and 50–80% moisture. White fish, such as cod, for example, contain roughly 20% protein, 80% water, 0.5–3% oil, minerals, vitamins and other components. Oily fish, such as mackerel and salmon, on the other hand, contain 20% protein, 10–18% oil and 62–70% water. The fish processing industry considers between 20% and 80% of fish to be waste, depending on numerous factors such as the type of fish and the processing standards. Head, viscera, skin, bones and scales account for 9–12%, 12–18%, 1–3%, 9–15%, and 5% of the total weight of the fish, respectively (Obiero et al., 2019). Recent research has revealed that such waste can be regarded as a valuable by-product source of value-added compounds. As a result, the minerals and bioactive substances found in fish by-products have received a lot of attention. For the pharmaceutical, nutraceutical and cosmeceutical industries, these materials are considered sustainable sources.

## 10.3. BIOACTIVE COMPOUNDS

Fish bioactive compounds are biologically active chemicals found in fish by-products. These components are good for human health. Multiple biological activities, such as antioxidant activity, hormone regulation, immune system enhancement, and facilitation of substance transition through the digestive tract, butyric acid production in the colon, and absorption and/or dilution of substances in the gut, contribute to these health benefits. As a result, fish by-products are a good source of bioactive chemicals that can be employed as dietary supplements and have medicinal and physiological benefits. Collagen is a fibrous and structural protein found in the extracellular space of fish that helps tissues such as bones, tendons, skin, cartilage and muscle function properly. It is the most abundant single protein in fish, accounting for 25% of the total protein content. Although there are several varieties of collagen, the most prevalent form found in fish by-products is Collagen type I, which is found in connective tissues, skin, muscles, bone and cornea. Collagen has been extracted from the skin of various fish species (Atef & Mahdi Ojagh, 2017). Some collagen-derived peptides, in particular, have been shown to have antioxidant activity, antimicrobial activity against various bacteria strains, and potent antihypertensive activity via ACE inhibitory properties. Gelatin is a proteinaceous macromolecule made from collagen that has been thermally denatured in a kinetically irreversible process. It can thus be used to improve food consistency, elasticity and stability as well as to create edible and biodegradable films that extend the shelf life of food. Furthermore, fish gelatin was found to have stronger antioxidant activity than synthetic gelatin. Gelatin was recovered from fish by-products in several experiments. However, fish skin was the predominant source of gelatin. It was isolated from the skin of seabass and pacific cod. Gelatin was also recovered from the bones of black tilapia, the scales of bighead carp and the head of mackerel.

## 10.4. PHENOLIC COMPOUNDS

Phenolic acids, flavonoids, catechins, tannins and phlorotannins are some of the phenolic chemicals found in marine species. The antioxidative action of phenolic compounds is the most essential role they play. The amount of total phenol compounds extracted from various marine species is inversely proportional to the extraction method used. Acetone 70% is the preferred solvent for extracting phenolic chemicals. After phenols have been extracted, purification procedures such as Silica gel chromatography, high pressure liquid chromatography (HPLC) and nuclear magnetic resonance (NMR) can be used. Plants, algae and fungi produce carotenoids as a by-product of photosynthesis. Different algal species have different types of carotenoids. Various carotenoids, such as carotene, violaxanthin, lutein, zeaxanthin, neoxanthin, astaxanthin and fucoxanthin have antioxidant, anti-obesity, antidiabetic, and anticancer properties. Carotenoids are widely used in the food business as colorants in products, in addition to their health benefits. Astaxanthin, zeaxanthin and lutein are the most frequent colored carotenoids used in food production. They have a wide range of applications in food. Solvent extraction with a nonpolar substance such as hexane is the usual method for obtaining algal carotenoids. When compared to traditional methods, new alternative approaches such as supercritical fluid extraction, pressurized liquid extraction, microwaves and ultrasound-assisted methods sped up the extraction process while lowering the amount of required solvent and energy consumption (Maqsood, Benjakul, & Shahidi, 2013).

## 10.5. CONSUMPTION

Dietary consumption has shifted dramatically in recent decades as a result of a variety of reasons, including the relationship between food composition and consumer health. Food additives were introduced into a complicated setting in which potential adverse effects were linked to their use, prompting researchers and food industry professionals to seek out more acceptable alternatives to standard food additives. Natural preservatives, antioxidants and colorants are attracting a lot of attention in this setting. Furthermore, quality degradation and eventual rejection of food products, whether due to loss of expected characteristics, formation of unpleasant attributes or for safety reasons, is dependent on each product, processing conditions, and storage conditions, necessitating the use of natural additives to improve quality and extend shelf life. Exploring aquaculture and by-products as sources of high-added value chemicals to make natural food additives could be a promising option to replace conventional food additives in this scenario. Each year, aquaculture produces a number of items that are not fit for human consumption. These by-products are proportional to the number of fish caught, accounting for 30% to 85% of the total weight. Fins, fishbones, gills, heads, liver, skin, viscera and other by-products from fisheries that are not intended for human consumption are among the applications for utilization in other purposes than for human consumption. Bioactive substances are phytochemicals that can be isolated from foods or food by-products and have the ability to control metabolic activities, resulting in positive effects. Aquaculture by-products contain a large number of bioactive chemicals. Among the most valuable molecules include collagen, chitin, enzymes, gelatin, glycosaminoglycans, polyunsaturated fatty acids (PUFA), minerals, protein, peptides and vitamins. These chemicals are said to have a number of nutritional benefits, which, when combined with their biological activity, could have a positive impact on human health. Some of the favorable bioactivities include anticancer, antidepressant, antihyperglycemic, antihypertensive, anti-inflammatory, antimicrobial, antioxidant, antiproliferative and immunomodulatory effects.

Once the desired valuable and helpful chemicals have been isolated and identified, a thorough evaluation of their functions should be carried out both before and after inclusion into the food matrix. Furthermore, in order to ensure the beneficial effects of bioactive compounds, it is necessary to understand how the digestion process may affect these compounds and the impact this may have on their stability, because the absorption process necessitates the release of bioactive compounds from the food matrix and their modification in the gastrointestinal tract through digestion. This includes changes to important molecules that could compromise their stability.

The bioavailability and bioactivity of bioactive substances will be determined as a result of such alterations.

These effects may be influenced by the gut microbiota and may vary from person to person. Bioavailability is the percentage of a bioactive component that is absorbed and utilized by the body, and hence determines the nutritional efficiency of that compound. Bioavailability refers to all of the events that determine a bioactive compound's nutritional efficiency in the body. This notion includes bioaccessibility, which is defined as the amount of a molecule liberated from the food matrix and modified in the gastrointestinal tract before becoming available for absorption.

## 10.6. CHRONIC DISORDERS

Chronic disorders such as obesity, diabetes, hypertension, hypercholesterolemia, cancer and cardiovascular diseases are becoming more prevalent as a result of major changes in the western lifestyle. Aquaculture by-products are produced in vast quantities and their bioactive chemicals can be profitably used for health or food supplements. Improving waste utilization in the aquaculture sector is crucial for a long-term industry that avoids or reduces environmental impact. In this regard, fish by-products are an excellent source of protein and omega-3 PUFA, which have been examined in Atlantic salmon and rainbow trout in particular. Fish protein hydrolysate (FPH) is a source of amino acids and peptides with high digestibility, rapid absorption and major biological activity derived by chemical, enzymatical and microbiological hydrolysis of processing by-products. Fish containing omega-3 PUFA like as eicosapentaenoic (EPA) and docosahexaenoic (DHA) have been shown to lower postprandial triacylglycerol levels, blood pressure, platelet aggregation and the inflammatory response. Crustacean by-products can also be utilized to make chitosan, which has antioxidant and antibacterial properties, and carotenoids, which have vital biological properties. Bioactive chemicals found in seaweeds include alginate, carrageenan, agar, carotenoids and polyphenols, which have antioxidant, anticancer, antidiabetic, antibacterial and anti-inflammatory properties. Finally, microalgae have primarily been used in the healthy food market in recent decades, with >75% of annual microalgal biomass production going into the creation of powders, tablets, capsules or pills.

Fish availability is important because of its nutritional value, which includes high levels of long-chain PUFA, protein, vital amino acids and minerals, all of which contribute to human health, which is especially important in low-income, food-insecure nations. Nonetheless, the fish market has undergone intensification processes that have had an unfavorable influence, gradually reduced marine biodiversity and the number of aquatic organisms while increased fish farming. It is commonly understood that wild catch fisheries will not be sufficient to meet the nutritional needs of an ever-increasing global population. As a result, aquaculture has been identified as a possible alternative for preserving marine resources, and its popularity has grown rapidly in recent decades, with no signs of slowing down. For example, aquaculture currently accounts for half of all seafood consumed, and global aquaculture production has even overtaken beef production. This expansion was mostly attributed to a high level of technological innovation, with production techniques shifting from somewhat extensive to more intense. The ability to provide fish for future populations will be determined by a variety of factors, including climate-driven changes in ecosystem productivity, fisheries management effectiveness, and the ability to expand aquaculture while reducing its environmental impacts. Despite the fact that aquaculture was originally developed for environmental reasons, its rapid industrialization and activity has resulted in a number of issues, including eutrophication and the dispersion of undesirable compounds such as antibiotics, hormones, and biocides. To supply for the long-term, reducing the impact of the food sector on the environment of an ever-growing human population is a huge problem. Furthermore, to improve aquaculture's sustainability, the practice of integrated multi-trophic aquaculture (IMTA), which involves cocultivation fed species such as finfish or shrimp with extractive species, which can feed on the various effluents produced by the aforementioned fed species, should be considered. This trend should be kept in mind at all levels of the manufacturing process, with new clean technologies and best practices being implemented. This fact has the potential to alter consumer perceptions of what is purchased and consumed, as well as the environmental and societal consequences.

The aquaculture sector faces issues that are difficult to solve, due in part to divergent viewpoints among industry, community and government stakeholders. Overall, aquaculture is an industrial activity that attempts to not only provide high-quality food, but also to replenish fish populations and preserve the biodiversity of our oceans. When these aims are considered, maintaining fish welfare is critical to ensuring the production of animals in similar conditions to their wild counterparts. In this regard, the conditions in which animals are bred must be constantly monitored in order to avoid pain or physical harm. All food production techniques, including aquaculture, require adequate regulatory requirements; otherwise, negative consequences for both the environment and industry are possible, such as the oceans' inability to meet rising demand. A crucial point is to have governance that is capable of regulating and managing the sector. Countries, on the whole, lack defined frameworks for supporting rising aquaculture growth, particularly in the case of offshore production.

Governments have suggested eating more fish based on the theory that omega-3 fatty acids found in fish oils are beneficial to human health by lowering the risk of cerebrovascular and other disorders. While European consumers are cautious of items generated from fish farming, the European Commission is working to instruct consumers about the benefits of this system and encourages them to accept aquaculture products. Indeed, aquaculture, when combined with the appropriate policy and fishery management actions, is required to ensure the long-term availability of fish as a vital source of human nutrition. Governments must remember that environmentally sustainable agriculture is critical in order to preserve the productivity and diversity of the food resource and the ecosystems that support it, as well as to avoid pressures and over-exploitation compromising other ecosystem functions. After a few years, the EU attempted to encourage the aquaculture sector through the revision of the Common Fisheries Policy (CFP) in 2013 and the associated Blue Growth agenda and produced a message consisting of Strategic Guidelines with common priorities and general goals.

In particular, the European Commission has invested significant sums, in collaboration with European Union countries, to develop the aquaculture sector and make it more competitive and successful by focusing on quality and safety, as well as environmentally friendly production. Specifically, the EU spent €1.17 billion on aquaculture between 2000 and 2014. The EU wants to invest an additional €1.72 billion on this activity between 2014 and 2020 through the European Maritime and Fisheries Fund (Guillen et al., 2019). The EU, on the other hand, may have failed to utilize the funds efficiently, which may have resulted in better results. Under the EU plan, funds are managed on a national level, with a wide range of options for implementing various types of projects. As a result, the majority of the subsidies go to small-scale initiatives. Despite all of the assistance, global output has expanded tremendously, EU production has remained unchanged. As a result, the EU is reliant on seafood imports. This situation necessitates the creation of better rules, as well as greater implementation of EU plans by member states. To ensure that most marginal groups may live in dignity and meet their most basic requirements, all governments must respond correctly to periods of food shortages and excessive price increases. The aquaculture business is being developed in a variety of methods around the world, including economic, social, and environmental development. Most modern regulations are similar to those in the EU, and they pose similar issues. Furthermore, governments from many countries, international organizations and private investors all promote creative advances. Fishing and aquaculture activities can serve a variety of purposes, ranging from meeting basic requirements to generating large revenue for multinational corporations operating in export markets. In this respect, the global impact of increased food consumption will differ by country. Capture output has declined in Europe, for example, but this has not been fully matched by a growth in aquaculture. In Asia, on the other hand, a gradual increase in captures was accompanied by a significant growth in aquaculture production. Asia produces by far the greatest amount of worldwide aquaculture production.

Provincial governments, non-governmental organizations, industrial partnerships and scientists are all participating in a war to gain public trust. Canada had the highest proportion of negative attitudes toward all types of aquacultures, as well as the most polarized split between positive and negative sentiments. A vital part of sustaining fish welfare is having adequate knowledge of the biology of farmed species, as well as the proper tools and technology to offer an acceptable environment and diet. Fish feeding production is influenced by the rise of aquaculture and rising demand, which has resulted in a shortage, high prices, and reduced ecological safety of traditional raw materials such as fish meal and fish oil. Wild-caught fishery products are commonly utilized in aquaculture feeds, which is why diverse aquaculture sectors are regarded as fish consumers rather than producers. By introducing technical attributes or specialized nutrients and bioactive components, using different options as raw materials could boost the value of the product. Among the unconventional sources of nutrients investigated in recent years, microbial biomass derived from a variety of microorganisms has been reported as a potential substitute for plant and animal derived ingredients, meeting protein and energy requirements while also adding functional properties.

On the other hand, the usage of insect meal as a sustainable alternative to fish meal has lately increased. Insects are simple to grow, effective at turning feed into biomass, have a low environmental impact, and have a high protein content. Furthermore, from a regulatory standpoint, the use of insect meals as food for aquaculture species has lately been recognized, unlike animal by-products, which have restrictions in terms of preventing human infections. Microalgae can boost the nutritional value of animal feed and play an important role in aquaculture. Microalgae, in fact, serve an important function in aquatic organism's physiological growth and external appearance. In an increasingly competitive market, several food certifications formats and labeling schemes that encourage sustainable and high-quality products are becoming important strategies to explore. For example, farmed seafood producers use product differentiation to satisfy consumer expectations and increase profits, taking into consideration current consumer tastes, which affect all stakeholders in the value chain. Labels assist consumers in selecting items that are consistent with their food attitudes. Government labels are almost always required and provide basic information on the most important attributes. Meanwhile, additional labels are usually optional and provide information about manufacturing procedures or key attributes of the product. The EU's efforts to strengthen and defend the aquaculture industry have boosted aquaculture product labeling policies. These include the requirement to identify the nation of origin, the opportunity to include information on production techniques, such as the Aquaculture Stewardship Council (ASC) eco-label, and the inclusion of nutrition and health information. Labels attempt to promote ecologically friendly, sustainable production, boost customer awareness, and urge a shift in purchasing habits.

The volume of fishing catch has reached a point of exploitation that poses a threat to marine species, necessitating the imposition of limits in order to prevent the extinction of marine biodiversity. Aquaculture alternatives, on the other hand, pose a concern to the environment, as they produce tons of phosphate and nitrogen. The build-up of feces around the breeding cages endangers marine resources and creates dead zones, preventing life growth. The public's perceptions of aquaculture's economic, social and environmental components can have a significant impact on industry practices and regulatory institutions. Consumer understanding of aquaculture and its products, on the other hand, is minimal, and many of them show only a limited interest in learning more. Similarly, the country-of-origin label is an important feature for buyers. Higher animal welfare and lower environmental impacts of fisheries and aquaculture have been cited as major drivers of consumer fish product choice in this regard. Consumer thoughts about how to proceed while making product-specific evaluations and purchasing decisions are frequently influenced by ethical beliefs. Advances in aquaculture should not jeopardize the appropriate working conditions for the persons involved, both socially and economically. Because of the growing need to reduce pressure on marine resources and because fish-derived products are essential in our diet, providing essential amino acids, high-quality animal protein, omega-3 fatty acids, minerals, vitamins, and other trace elements, the European Commission is urging consumers to accept aquaculture products on a broad scale.

Similarly, the absence of approved veterinary medicines and vaccinations is a significant issue for the industry. The environmental impact assessment is another issue in aquaculture legislation (EIA). In actuality, there are significant differences in EIA methodologies not only among member states, but also within a country's internal extension, due to fragmented public entities with autonomy and a range of legislation. As a result, there are no standard protocols for conducting EIAs for aquaculture projects, and they take a long time to complete. Furthermore, policymakers will need to be more proactive in educating the public about the industry's major contemporary implications. The production of thorough information to assist people in better understanding the difficulties surrounding the aquaculture business may lead to increased acceptability. It's unsurprising that the involvement of several institutions can result in policy, regulation, administration and control conflicts or overlaps. Several studies have pointed to the potential advantages of diversifying human diets to include protein sources other than meat, such as seafood. In fact, for the time being, and for the foreseeable future, the majority of seafood is produced from the farming business. Aquaculture technology currently in use is a viable technique for reducing some negative effects, such as pollution and disease.

# BIBLIOGRAPHY


1. AbuDalo, M., El-khateeb, M., Ayadi, H., Al-Rahahleh, B., Jaradat, A., & Guermazi, W., (2021). First assessment of water quality of an artificial lake for fish culture and irrigation: A case study of water reuse in water shortage area across the Middle East. *Aquaculture Research, 52*(3), 1267–1281. https://doi.org/10.1111/are.14985.
2. Adeleke, B., Robertson-Andersson, D., Moodley, G., & Taylor, S., (2021). Aquaculture in Africa: A comparative review of Egypt, Nigeria, and Uganda Vis-À-Vis South Africa. *Reviews in Fisheries Science & Aquaculture, 29*(2), 167–197. https://doi.org/10.1080/23308249.2020.1795615.
3. Ahmad, A., Sheikh, A. S. R., Hasan, H. A., Othman, A. R., & Ismail, N. I., (2021). Aquaculture industry: Supply and demand, best practices, effluent and its current issues and treatment technology. *Journal of Environmental Management, 287*, 112271. https://doi.org/10.1016/j.jenvman.2021.112271.
4. Ahmed, N., & Turchini, G. M., (2021). Recirculating aquaculture systems (RAS): Environmental solution and climate change adaptation. *Journal of Cleaner Production, 297*, 126604. https://doi.org/10.1016/j.jclepro.2021.126604.
5. Amekawa, Y., (2009). Reflections on the growing influence of good agricultural practices in the global South. *Journal of Agricultural and*

*Environmental Ethics, 22*(6), 531. https://doi.org/10.1007/s10806-009-9171-8.

6. Amlund, H., Berntssen, M. H. G., Lunestad, B. T., & Lundebye, A. K., (2012). 10—Aquaculture feed contamination by persistent organic pollutants, heavy metals, additives and drug residues. In: Fink-Gremmels, J., (ed.), *Animal Feed Contamination* (pp. 205–229). Woodhead Publishing. https://doi.org/10.1533/9780857093615.2.205.
7. Arlinghaus, R., Mehner, T., & Cowx, I. G., (2002). Reconciling traditional inland fisheries management and sustainability in industrialized countries, with emphasis on Europe. *Fish and Fisheries, 3*(4), 261–316. https://doi.org/10.1046/j.1467-2979.2002.00102.x.
8. Asche, F., Roll, K. H., & Trollvik, T., (2009). New aquaculture species—The whitefish market. *Aquaculture Economics & Management, 13*(2), 76–93. https://doi.org/10.1080/13657300902881641.
9. Atef, M., & Mahdi, O. S., (2017). Health benefits and food applications of bioactive compounds from fish byproducts: A review. *Journal of Functional Foods, 35*, 673–681. https://doi.org/10.1016/j.jff.2017.06.034.
10. Axler, R., Larsen, C., Tikkanen, C., McDonald, M., Yokom, S., & Aas, P., (1996). Water quality issues associated with aquaculture: A case study in mine pit lakes. *Water Environment Research, 68*(6), 995–1011. https://doi.org/10.2175/106143096X128027.
11. Azim, M. E., Wahab, M. A., Biswas, P. K., Asaeda, T., Fujino, T., & Verdegem, M. C. J., (2004). The effect of periphyton substrate density on production in freshwater polyculture ponds. *Aquaculture, 232*(1), 441–453. https://doi.org/10.1016/j.aquaculture.2003.08.010.
12. Baganz, G. F. M., Junge, R., Portella, M. C., Goddek, S., Keesman, K. J., Baganz, D., & Kloas, W., (2022). The aquaponic principle-It is all about coupling. *Reviews in Aquaculture, 14*(1), 252–264. https://doi.org/10.1111/raq.12596.
13. Barlow, S. M., (2003). Fish meal. In: Caballero, B., (ed.), *Encyclopedia of Food Sciences and Nutrition* (2nd edn., pp. 2486–2491). Oxford: Academic Press. https://doi.org/10.1016/B0-12-227055-X/00479-X.
14. Barrut, B., Blancheton, J. P., Champagne, J. Y., & Grasmick, A., (2012). Mass transfer efficiency of a vacuum airlift—Application to water recycling in aquaculture systems. *Aquacultural Engineering, 46*, 18–26. https://doi.org/10.1016/j.aquaeng.2011.10.004.

15. Bartelme, R. P., McLellan, S. L., & Newton, R. J., (2017). Freshwater recirculating aquaculture system operations drive biofilter bacterial community shifts around a stable nitrifying consortium of ammonia-oxidizing *Archaea* and comammox *Nitrospira*. *Frontiers in Microbiology, 8*. Retrieved from: https://www.frontiersin.org/article/10.3389/fmicb.2017.00101 (accessed on 25 June 2022).
16. Barton, B. A., & Iwama, G. K., (1991). Physiological changes in fish from stress in aquaculture with emphasis on the response and effects of corticosteroids. *Annual Review of Fish Diseases, 1*, 3–26. https://doi.org/10.1016/0959-8030(91)90019-G.
17. Behroozi, L., & Couturier, M. F., (2019). Prediction of water velocities in circular aquaculture tanks using an axisymmetric CFD model. *Aquacultural Engineering, 85*, 114–128. https://doi.org/10.1016/j.aquaeng.2019.03.005.
18. Belton, B., & Azad, A., (2012). The characteristics and status of pond aquaculture in Bangladesh. *Aquaculture,* 358, 359, 196–204. https://doi.org/10.1016/j.aquaculture.2012.07.002.
19. Benetti, D. D., Leingang, A. J., Russo, R., Powell, T. M., Cleary, D., Grabe, S. W., & Main, K. L., (2001). Development of aquaculture methods for southern flounder, *Paralichthys lethostigma. Journal of Applied Aquaculture, 11*(1, 2), 135–146. https://doi.org/10.1300/J028v11n01_10.
20. Bess, R., (2006). New Zealand seafood firm competitiveness in export markets: The role of the quota management system and aquaculture legislation. *Marine Policy, 30*(4), 367–378. https://doi.org/10.1016/j.marpol.2005.06.011.
21. Bigger, C. H., Murray, K., & Murray, N. E., (1973). Recognition sequence of a restriction enzyme. *Nature New Biology, 244*(131), 7–10. https://doi.org/10.1038/newbio244007a0.
22. Bjørndal, T., Lane, D. E., & Weintraub, A., (2004). Operational research models and the management of fisheries and aquaculture: A review. *European Journal of Operational Research, 156*(3), 533–540. https://doi.org/10.1016/S0377-2217(03)00107-3.
23. Blanchard, C., (2019). *Improving Greenhouse Cucumber Production in De-coupled Aquaponic Systems*. Retrieved from: https://etd.auburn.edu//handle/10415/6958 (accessed on 25 June 2022).

24. Blancheton, J. P., Attramadal, K. J. K., Michaud, L., d'Orbcastel, E. R., & Vadstein, O., (2013). Insight into bacterial population in aquaculture systems and its implication. *Aquacultural Engineering, 53*, 30–39. https://doi.org/10.1016/j.aquaeng.2012.11.009.
25. Blidariu, F., & Grozea, A., (2011). Increasing the economical efficiency and sustainability of indoor fish farming by means of aquaponics-review. *Scientific Papers: Animal Science and Biotechnologies, 44*.
26. Bonar, S. A., & Hubert, W. A., (2002). Standard sampling of inland fish: Benefits, challenges, and a call for action. *Fisheries, 27*(3), 10–16. https://doi.org/10.1577/1548-8446(2002)027<0010:SSOIF>2.0.CO;2.
27. Boyd, C. E., & Pillai, V. K., (1985). *Water Quality Management in Aquaculture* (Vol. 22, pp. 1–44). CMFRI Special Publication. Retrieved from: http://eprints.cmfri.org.in/3264/ (accessed on 25 June 2022).
28. Bricknell, I. R., Birkel, S. D., Brawley, S. H., Van, K. T., Hamlin, H. J., Capistrant-Fossa, K., & Moeykens, S., (2021). Resilience of cold water aquaculture: A review of likely scenarios as climate changes in the Gulf of Maine. *Reviews in Aquaculture, 13*(1), 460–503. https://doi.org/10.1111/raq.12483.
29. Bueno, G. W., Ostrensky, A., Canzi, C., De Matos, F. T., & Roubach, R., (2015). Implementation of aquaculture parks in Federal Government waters in Brazil. *Reviews in Aquaculture, 7*(1), 1–12. https://doi.org/10.1111/raq.12045.
30. Cao, L., Wang, W., Yang, Y., Yang, C., Yuan, Z., Xiong, S., & Diana, J., (2007). Environmental impact of aquaculture and countermeasures to aquaculture pollution in China. *Environmental Science and Pollution Research-International, 14*(7), 452–462. https://doi.org/10.1065/espr2007.05.426.
31. Chatla, D., Padmavathi, P., & Srinu, G., (2020). Wastewater treatment techniques for sustainable aquaculture. In: Ghosh, S. K., (ed.), *Waste Management as Economic Industry Towards Circular Economy* (pp. 159–166). Singapore: Springer. https://doi.org/10.1007/978-981-15-1620-7_17.
32. Chen, H., Liu, S., Xu, X. R., Diao, Z. H., Sun, K. F., Hao, Q. W., & Ying, G. G. (2018). Tissue distribution, bioaccumulation characteristics and health risk of antibiotics in cultured fish from a typical aquaculture area. *Journal of Hazardous Materials, 343*, 140–148. https://doi.org/10.1016/j.jhazmat.2017.09.017.

33. Cho, C. Y., Hynes, J. D., Wood, K. R., & Yoshida, H. K., (1994). Development of high-nutrient-dense, low-pollution diets and prediction of aquaculture wastes using biological approaches. *Aquaculture, 124*(1), 293–305. https://doi.org/10.1016/0044-8486(94)90403-0.

34. Chua, T. E., Paw, J. N., & Guarin, F. Y., (1989). The environmental impact of aquaculture and the effects of pollution on coastal aquaculture development in Southeast Asia. *Marine Pollution Bulletin, 20*(7), 335–343. https://doi.org/10.1016/0025-326X(89)90157-4.

35. Costa-Pierce, B. A., (1987). Initial feasibility of integrated aquaculture on lava soils in Hawaii. *Aquacultural Engineering, 6*(3), 171–182. https://doi.org/10.1016/0144-8609(87)90002-1.

36. Danish, M. S. S., Senjyu, T., Sabory, N. R., Khosravy, M., Grilli, M. L., Mikhaylov, A., & Majidi, H., (2021). A forefront framework for sustainable aquaponics modeling and design. *Sustainability, 13*(16), 9313. https://doi.org/10.3390/su13169313.

37. Davidson, J., Helwig, N., & Summerfelt, S. T., (2008). Fluidized sand biofilters used to remove ammonia, biochemical oxygen demand, total coliform bacteria, and suspended solids from an intensive aquaculture effluent. *Aquacultural Engineering, 39*(1), 6–15. https://doi.org/10.1016/j.aquaeng.2008.04.002.

38. Defoirdt, T., (2014). Virulence mechanisms of bacterial aquaculture pathogens and antivirulence therapy for aquaculture. *Reviews in Aquaculture, 6*(2), 100–114. https://doi.org/10.1111/raq.12030.

39. Donaldson, E. M., (1986). The integrated development and application of controlled reproduction techniques in Pacific salmonid aquaculture. *Fish Physiology and Biochemistry, 2*(1), 9–24. https://doi.org/10.1007/BF02264070.

40. Enduta, A., Jusoh, A., Ali, N., & Wan, N. W. B., (2011). Nutrient removal from aquaculture wastewater by vegetable production in aquaponics recirculation system. *Desalination and Water Treatment, 32*(1–3), 422–430. https://doi.org/10.5004/dwt.2011.2761.

41. Everson, J. L., Jones, D. R., Taylor, A. K., Rutan, B. J., Leeds, T. D., Langwig, K. E., & Wiens, G. D., (2021). Aquaculture reuse water, genetic line, and vaccination affect rainbow trout (*Oncorhynchus mykiss*) disease susceptibility and infection dynamics. *Frontiers in Immunology, 12*, 721048. https://doi.org/10.3389/fimmu.2021.721048.

42. Flegel, T. W., (2019). A future vision for disease control in shrimp aquaculture. *Journal of the World Aquaculture Society, 50*(2), 249–266. https://doi.org/10.1111/jwas.12589.
43. Fletcher, Z. J., (2022). Aquaculture in multiple use of space for island clean autonomy. In: Piątek, L., Lim, S. H., Wang, C. M., & De Graaf-Van, D. R., (eds.), *WCFS2020* (pp. 193–200). Singapore: Springer. https://doi.org/10.1007/978-981-16-2256-4_12.
44. Føre, M., Frank, K., Norton, T., Svendsen, E., Alfredsen, J. A., Dempster, T., & Berckmans, D., (2018). Precision fish farming: A new framework to improve production in aquaculture. *Biosystems Engineering, 173*, 176–193. https://doi.org/10.1016/j.biosystemseng.2017.10.014.
45. Froehlich, H. E., Gentry, R. R., Lester, S. E., Cottrell, R. S., Fay, G., Branch, T. A., & Baum, J. K., (2021). Securing a sustainable future for US seafood in the wake of a global crisis. *Marine Policy, 124*, 104328. https://doi.org/10.1016/j.marpol.2020.104328.
46. Gaikowski, M. P., Larson, W. J., Steuer, J. J., & Gingerich, W. H., (2004). Validation of two dilution models to predict chloramine-T concentrations in aquaculture facility effluent. *Aquacultural Engineering, 30*(3), 127–140. https://doi.org/10.1016/j.aquaeng.2003.11.001.
47. Gentry, R. R., Lester, S. E., Kappel, C. V., White, C., Bell, T. W., Stevens, J., & Gaines, S. D., (2017). Offshore aquaculture: Spatial planning principles for sustainable development. *Ecology and Evolution, 7*(2), 733–743. https://doi.org/10.1002/ece3.2637.
48. Ghittino, C., Latini, M., Agnetti, F., Panzieri, C., Lauro, L., Ciappelloni, R., & Petracca, G., (2003). Emerging pathologies in aquaculture: Effects on production and food safety. *Veterinary Research Communications, 27*(1), 471–479. https://doi.org/10.1023/B:VERC.0000014204.37722.b6.
49. Gladue, R. M., & Maxey, J. E., (1994). Microalgal feeds for aquaculture. *Journal of Applied Phycology, 6*(2), 131–141. https://doi.org/10.1007/BF02186067.
50. Gregory, S. P., Dyson, P. J., Fletcher, D., Gatland, P., & Shields, R. J., (2012). Nitrogen removal and changes to microbial communities in model flood/drain and submerged biofilters treating aquaculture wastewater. *Aquacultural Engineering, 50*, 37–45. https://doi.org/10.1016/j.aquaeng.2012.03.006.
51. Guillen, J., Asche, F., Carvalho, N., Fernández, P. J. M., Llorente, I., Nielsen, R., & Villasante, S., (2019). Aquaculture subsidies

in the European Union: Evolution, impact and future potential for growth. *Marine Policy, 104*, 19–28. https://doi.org/10.1016/j.marpol.2019.02.045.

52. Gullian, M., Espinosa-Faller, F. J., Núñez, A., & López-Barahona, N., (2012). Effect of turbidity on the ultraviolet disinfection performance in recirculating aquaculture systems with low water exchange. *Aquaculture Research, 43*(4), 595–606. https://doi.org/10.1111/j.1365-2109.2011.02866.x.
53. Hargreaves, J. A., (2006). Photosynthetic suspended-growth systems in aquaculture. *Aquacultural Engineering, 34*(3), 344–363. https://doi.org/10.1016/j.aquaeng.2005.08.009.
54. Hasan, Z., & Andriani, Y., (2018). Novel mechanical filter for reducing ammonia concentration of silver barb culture in a recirculating aquaculture system (RAS). *Research Journal of Chemistry and Environment, 22*, 319–324.
55. Holtsmark, M., Klemetsdal, G., Sonesson, A. K., & Woolliams, J. A., (2008). Establishing a base population for a breeding program in aquaculture, from multiple subpopulations, differentiated by genetic drift: I. Effects of the number of subpopulations, heritability and mating strategies using optimum contribution selection. *Aquaculture, 274*(2), 232–240. https://doi.org/10.1016/j.aquaculture.2007.11.042.
56. Huang, C. T., Chiou, J. T., Khac, H. T., Hsiao, Y. J., & Chen, S. C. (2016). Improving the management of commercial giant mottled eel *Anguilla marmorata* aquaculture in Taiwan for improved productivity: A bioeconomic analysis. *Fisheries Science, 82*(1), 95–111. https://doi.org/10.1007/s12562-015-0934-z.
57. Huggins, D. L., Piedrahita, R. H., & Rumsey, T., (2004). Analysis of sediment transport modeling using computational fluid dynamics (CFD) for aquaculture raceways. *Aquacultural Engineering, 31*(3), 277–293. https://doi.org/10.1016/j.aquaeng.2004.05.007.
58. Jacobs, M. N., Covaci, A., & Schepens, P., (2002). Investigation of selected persistent organic pollutants in farmed Atlantic salmon (*Salmo salar*), salmon aquaculture feed, and fish oil components of the feed. *Environmental Science & Technology, 36*(13), 2797–2805. https://doi.org/10.1021/es011287i.
59. Jasmin, M. Y., Syukri, F., Kamarudin, M. S., & Karim, M., (2020). Potential of bioremediation in treating aquaculture sludge: Review

article. *Aquaculture, 519*, 734905. https://doi.org/10.1016/j.aquaculture.2019.734905.

60. Jensen, M. H., (1999). Hydroponics worldwide. *Acta Horticulturae*, (481), 719–730. https://doi.org/10.17660/ActaHortic.1999.481.87.
61. Johnson, P. W., & Sieburth, J. M., (1976). In situ morphology of nitrifying-like bacteria in aquaculture systems. *Applied and Environmental Microbiology, 31*(3), 423–432. https://doi.org/10.1128/aem.31.3.423-432.1976.
62. Kawarazuka, N., (2010). *The Contribution of Fish Intake, Aquaculture, and Small-Scale Fisheries to Improving Nutrition Security: A Literature Review*. http://Aquaticcommons.Org/Id/Eprint/3469. Retrieved from https://aquadocs.org/handle/1834/21201 (accessed on 25 June 2022).
63. Kawarazuka, N., & Béné, C., (2010). Linking small-scale fisheries and aquaculture to household nutritional security: An overview. *Food Security, 2*(4), 343–357. https://doi.org/10.1007/s12571-010-0079-y.
64. Kuhn, D. D., Drahos, D. D., Marsh, L., & Flick, G. J., (2010). Evaluation of nitrifying bacteria product to improve nitrification efficacy in recirculating aquaculture systems. *Aquacultural Engineering, 43*(2), 78–82. https://doi.org/10.1016/j.aquaeng.2010.07.001.
65. Labatut, R. A., Ebeling, J. M., Bhaskaran, R., & Timmons, M. B., (2007). Hydrodynamics of a large-scale mixed-cell raceway (MCR): Experimental studies. *Aquacultural Engineering, 37*(2), 132–143. https://doi.org/10.1016/j.aquaeng.2007.04.001.
66. Lai, H. T., & Lin, J. J., (2009). Degradation of oxolinic acid and flumequine in aquaculture pond waters and sediments. *Chemosphere, 75*(4), 462–468. https://doi.org/10.1016/j.chemosphere.2008.12.060.
67. Le Gouvello, R., Hochart, L. E., Laffoley, D., Simard, F., Andrade, C., Angel, D., & Marino, G., (2017). Aquaculture and marine protected areas: Potential opportunities and synergies. *Aquatic Conservation: Marine and Freshwater Ecosystems, 27*(S1), 138–150. https://doi.org/10.1002/aqc.2821.
68. Li, D., Wang, Z., Wu, S., Miao, Z., Du, L., & Duan, Y., (2020). Automatic recognition methods of fish feeding behavior in aquaculture: A review. *Aquaculture, 528*, 735508. https://doi.org/10.1016/j.aquaculture.2020.735508.
69. Li, P., Wang, C., Liu, G., Luo, X., Rauan, A., Zhang, C., & Gao, Q., (2022). A hydroponic plants and biofilm combined treatment system

efficiently purified wastewater from cold flowing water aquaculture. *Science of The Total Environment, 821*, 153534. https://doi.org/10.1016/j.scitotenv.2022.153534.

70. Liu, J., & Cai, Q., (1998). Integrated aquaculture in Chinese lakes and paddy fields. Paper presented at ICEE—international conference on ecological engineering, Beijing, China. *Ecological Engineering, 11*(1), 49–59. https://doi.org/10.1016/S0925-8574(98)00022-6.
71. Love, D. C., Fry, J. P., Li, X., Hill, E. S., Genello, L., Semmens, K., & Thompson, R. E., (2015). Commercial aquaponics production and profitability: Findings from an international survey. *Aquaculture, 435*, 67–74. https://doi.org/10.1016/j.aquaculture.2014.09.023.
72. Loyless, J. C., & Malone, R. F., (1998). Evaluation of air-lift pump capabilities for water delivery, aeration, and degasification for application to recirculating aquaculture systems. *Aquacultural Engineering, 18*(2), 117–133. https://doi.org/10.1016/S0144-8609(98)00025-9.
73. MacMillan, J. R., Huddleston, T., Woolley, M., & Fothergill, K., (2003). Best management practice development to minimize environmental impact from large flow-through trout farms. *Aquaculture, 226*(1), 91–99. https://doi.org/10.1016/S0044-8486(03)00470-8.
74. Mahmudov, K., Mahmoud, A., Sur, S., Cruz, F. C., & Bilton, A. M., (2019). Feasibility of a wind-powered aeration system for small-scale aquaculture in developing countries. *Energy for Sustainable Development, 51*, 40–49. https://doi.org/10.1016/j.esd.2019.05.003.
75. Maqsood, S., Benjakul, S., & Shahidi, F., (2013). Emerging role of phenolic compounds as natural food additives in fish and fish products. *Critical Reviews in Food Science and Nutrition, 53*(2), 162–179. https://doi.org/10.1080/10408398.2010.518775.
76. Martínez-Córdova, L. R., Martínez-Porchas, M., Emerenciano, M. G. C., Miranda-Baeza, A., & Gollas-Galván, T., (2017). From microbes to fish the next revolution in food production. *Critical Reviews in Biotechnology, 37*(3), 287–295. https://doi.org/10.3109/07388551.2016.1144043.
77. Martins, C., Eding, E., Verdegem, M., Heinsbroek, L. T. N., Schneider, O., Blancheton, J. P., & Verreth, J., (2010). New developments in recirculating aquaculture systems in Europe: A perspective on environmental sustainability. *Aquacultural Engineering (0144-8609)* (Vol. 43, No. 3, pp. 83–93). Elsevier Sci Ltd. https://doi.org/10.1016/j.aquaeng.2010.09.002.

78. Masabni, J., & Niu, G., (2022). Chapter 10—Aquaponics. In: Kozai, T., Niu, G., & Masabni, J., (eds.), *Plant Factory Basics, Applications and Advances* (pp. 167–180). Academic Press. https://doi.org/10.1016/B978-0-323-85152-7.00017-3.
79. Masci, M., Orban, E., & Nevigato, T., (2014). Organochlorine pesticide residues: An extensive monitoring of Italian fishery and aquaculture. *Chemosphere, 94*, 190–198. https://doi.org/10.1016/j.chemosphere.2013.10.016.
80. Mohapatra, B., Sarkar, B., Barik, N., & P, J., (2011). *Application of Plastics in Aquaculture*. 1, 25-50.
81. Montes, C., Acharya, N., Rumana, H. P., Amjath, B. T. S., Krupnik, T. J., & Quamrul, H. S. M., (2022). Developing a framework for an early warning system of seasonal temperature and rainfall tailored to aquaculture in Bangladesh. *Climate Services, 26*, 100292. https://doi.org/10.1016/j.cliser.2022.100292.
82. Morrow, J. C., Reilly, J. J., Olivas, J. P., & Romero, M. M., (2001). Inexpensive Pipe-hangers, clamps, and braces for aquaculture applications. *North American Journal of Aquaculture, 63*(1), 79–82. https://doi.org/10.1577/1548-8454(2001)063<0079:IPHCAB>2.0.CO;2.
83. Mugimba, K. K., Byarugaba, D. K., Mutoloki, S., Evensen, Ø., & Munang'andu, H. M., (2021). Challenges and solutions to viral diseases of finfish in marine aquaculture. *Pathogens, 10*(6), 673. https://doi.org/10.3390/pathogens10060673.
84. Ndanga, L. Z. B., Quagrainie, K. K., & Dennis, J. H., (2013). Economically feasible options for increased women participation in Kenyan aquaculture value chain. *Aquaculture,* 414, 415, 183–190. https://doi.org/10.1016/j.aquaculture.2013.08.012.
85. Njoku, D. C., Allison, M. E., & Jegede, O. I., (2009). The characteristics of the moonfish, citharinus citharus (Eoffroy–St Hilaire, 1809) of Lake Oguta, Nigeria. *Animal Production Research Advances, 5*(1). https://doi.org/10.4314/apra.v5i1.49817.
86. O'Reilly, P., & Wright, J. M., (1995). The evolving technology of DNA fingerprinting and its application to fisheries and aquaculture. *Journal of Fish Biology, 47*(sA), 29–55. https://doi.org/10.1111/j.1095-8649.1995.tb06042.x.
87. Obiero, K., Meulenbroek, P., Drexler, S., Dagne, A., Akoll, P., Odong, R., & Waidbacher, H., (2019). The contribution of fish to

food and nutrition security in Eastern Africa: Emerging trends and future outlooks. *Sustainability, 11*(6), 1636. https://doi.org/10.3390/su11061636.

88. Olsen, R. L., & Hasan, M. R., (2012). A limited supply of fishmeal: Impact on future increases in global aquaculture production. *Trends in Food Science & Technology, 27*(2), 120–128. https://doi.org/10.1016/j.tifs.2012.06.003.
89. Pankhurst, N. W., & King, H. R., (2010). Temperature and salmonid reproduction: Implications for aquaculture. *Journal of Fish Biology, 76*(1), 69–85. https://doi.org/10.1111/j.1095-8649.2009.02484.x.
90. Paperna, I., (1991). Diseases caused by parasites in the aquaculture of warm water fish. *Annual Review of Fish Diseases, 1*, 155–194. https://doi.org/10.1016/0959-8030(91)90028-I.
91. Parker, N. C., & Suttle, M. A., (1987). Design of airlift pumps for water circulation and aeration in aquaculture. *Aquacultural Engineering, 6*(2), 97–110. https://doi.org/10.1016/0144-8609(87)90008-2.
92. Pedersen, S., & Wik, T., (2020). A comparison of topologies in recirculating aquaculture systems using simulation and optimization. *Aquacultural Engineering, 89*, 102059. https://doi.org/10.1016/j.aquaeng.2020.102059.
93. Philipose, M. T., (1966). *Present Trends in the Control of Weeds in Fish Cultural Waters of Asia and the Far East.* Retrieved from: https://www.cabdirect.org/cabdirect/abstract/19682302230 (accessed on 25 June 2022).
94. Publication: USDA ARS. (n.d.). Retrieved from https://www.ars.usda.gov/research/publications/publication/?seqNo115=303721 (accessed on 25 June 2022).
95. Rasdi, N. W., & Qin, J. G., (2016). Improvement of copepod nutritional quality as live food for aquaculture: A review. *Aquaculture Research, 47*(1), 1–20. https://doi.org/10.1111/are.12471.
96. Reid, G. K., Liutkus, M., Robinson, S. M. C., Chopin, T. R., Blair, T., Lander, T., & Moccia, R. D., (2009). A review of the biophysical properties of salmonid faeces: Implications for aquaculture waste dispersal models and integrated multi-trophic aquaculture. *Aquaculture Research, 40*(3), 257–273. https://doi.org/10.1111/j.1365-2109.2008.02065.x.

97. Roleda, M. Y., & Hurd, C. L., (2019). Seaweed nutrient physiology: Application of concepts to aquaculture and bioremediation. *Phycologia, 58*(5), 552–562. https://doi.org/10.1080/00318884.2019.1622920.

98. Rosa, J., Lemos, M. F. L., Crespo, D., Nunes, M., Freitas, A., Ramos, F., & Leston, S., (2020). Integrated multitrophic aquaculture systems–potential risks for food safety. *Trends in Food Science & Technology, 96*, 79–90. https://doi.org/10.1016/j.tifs.2019.12.008.

99. Rummer, J. L., & Brauner, C. J., (2011). Plasma-accessible carbonic anhydrase at the tissue of a teleost fish may greatly enhance oxygen delivery: *In vitro* evidence in rainbow trout, *Oncorhynchus mykiss*. *Journal of Experimental Biology, 214*(14), 2319–2328. https://doi.org/10.1242/jeb.054049.

100. Sánchez-Mata, A., & Mora, J., (2000). A review of marine aquaculture in Spain: Production, regulations and environmental monitoring. *Journal of Applied Ichthyology, 16*(4, 5), 209–213. https://doi.org/10.1046/j.1439-0426.2000.00261.x.

101. Sanni, S., & Forsberg, O. I., (1996). Modeling pH and carbon dioxide in single-pass sea-water aquaculture systems. *Aquacultural Engineering, 15*(2), 91–110. https://doi.org/10.1016/0144-8609(95)00003-8.

102. Scanes, C. G., (2018). Chapter 7-Animal agriculture: Livestock, poultry, and fish aquaculture. In: Scanes, C. G., & Toukhsati, S. R., (eds.), *Animals and Human Society* (pp. 133–179). Academic Press. https://doi.org/10.1016/B978-0-12-805247-1.00007-1.

103. Seawright, D. E., Stickney, R. R., & Walker, R. B., (1998). Nutrient dynamics in integrated aquaculture–hydroponics systems. *Aquaculture, 160*(3), 215–237. https://doi.org/10.1016/S0044-8486(97)00168-3.

104. Sepúlveda, F., Marín, S. L., & Carvajal, J., (2004). Metazoan parasites in wild fish and farmed salmon from aquaculture sites in southern Chile. *Aquaculture, 235*(1), 89–100. https://doi.org/10.1016/j.aquaculture.2003.09.015.

105. Shafer, C. S., Inglis, G. J., & Martin, V., (2010). Examining residents' proximity, recreational use, and perceptions regarding proposed aquaculture development. *Coastal Management, 38*(5), 559–574. https://doi.org/10.1080/08920753.2010.511700.

106. Siddiqui, S., Conkle, J. L., Scarpa, J., & Sadovski, A., (2020). An analysis of U.S. wastewater treatment plant effluent dilution ratio: Implications for water quality and aquaculture. *Science of The Total Environment, 721*, 137819. https://doi.org/10.1016/j.scitotenv.2020.137819.

107. Smith, D. W., (1988). Phytoplankton and catfish culture: A review. *Aquaculture, 74*(3), 167–189. https://doi.org/10.1016/0044-8486(88)90361-4.
108. Soliman, A. I. E., Morad, M. M., Wasfy, K. I., & Moursy, M. A. M., (2020). Utilization of aquaculture drainage for enhancing onion crop yield under surface and subsurface drip irrigation systems. *Agricultural Water Management, 239*, 106244. https://doi.org/10.1016/j.agwat.2020.106244.
109. Song, X., Liu, Y., Pettersen, J. B., Brandão, M., Ma, X., Røberg, S., & Frostell, B., (2019). Life cycle assessment of recirculating aquaculture systems: A case of Atlantic salmon farming in China. *Journal of Industrial Ecology, 23*(5), 1077–1086. https://doi.org/10.1111/jiec.12845.
110. Stucchi, D., Sutherland, T. A., Levings, C., & Higgs, D., (2005). Near-field depositional model for salmon aquaculture waste. In: Hargrave, B. T., (ed.), *Environmental Effects of Marine Finfish Aquaculture* (pp. 157–179). Berlin, Heidelberg: Springer. https://doi.org/10.1007/b136009.
111. Timmons, M. B., Chen, S., & Weeks, N. C., (1995). Mathematical model of a foam fractionator used in aquaculture. *Journal of the World Aquaculture Society, 26*(3), 225–233. https://doi.org/10.1111/j.1749-7345.1995.tb00250.x.
112. Timmons, M. B., Summerfelt, S. T., & Vinci, B. J., (1998). Review of circular tank technology and management. *Aquacultural Engineering, 18*(1), 51–69. https://doi.org/10.1016/S0144-8609(98)00023-5.
113. Tomasso, J. R., (1994). Toxicity of nitrogenous wastes to aquaculture animals. *Reviews in Fisheries Science, 2*(4), 291–314. https://doi.org/10.1080/10641269409388560.
114. Tovar, A., Moreno, C., Mánuel-Vez, M. P., & García-Vargas, M., (2000). Environmental implications of intensive marine aquaculture in earthen ponds. *Marine Pollution Bulletin, 40*(11), 981–988. https://doi.org/10.1016/S0025-326X(00)00040-0.
115. Tripathi, S., & Tripathi, B. D., (2011). Efficiency of combined process of ozone and bio-filtration in the treatment of secondary effluent. *Bioresource Technology, 102*(13), 6850–6856. https://doi.org/10.1016/j.biortech.2011.04.035.
116. Trottet, A., George, C., Drillet, G., & Lauro, F. M., (2022). Aquaculture in coastal urbanized areas: A comparative review of the challenges

posed by harmful algal blooms. *Critical Reviews in Environmental Science and Technology, 52*(16), 2888–2929. https://doi.org/10.1080/10643389.2021.1897372.

117. Van, R. J., (2013). Waste treatment in recirculating aquaculture systems. *Aquacultural Engineering, 53*, 49–56. https://doi.org/10.1016/j.aquaeng.2012.11.010.
118. Vergara-Solana, F., Araneda, M. E., & Ponce-Díaz, G., (2019). Opportunities for strengthening aquaculture industry through multicriteria decision-making. *Reviews in Aquaculture, 11*(1), 105–118. https://doi.org/10.1111/raq.12228.
119. Viadero, R. C., & Noblet, J. A., (2002). Membrane filtration for removal of fine solids from aquaculture process water. *Aquacultural Engineering, 26*(3), 151–169. https://doi.org/10.1016/S0144-8609(02)00011-0.
120. Vogt, G., (1987). Monitoring of environmental pollutants such as pesticides in prawn aquaculture by histological diagnosis. *Aquaculture, 67*(1), 157–164. https://doi.org/10.1016/0044-8486(87)90021-4.
121. Wietz, M., Hall, M. R., & Høj, L., (2009). Effects of seawater ozonation on biofilm development in aquaculture tanks. *Systematic and Applied Microbiology, 32*(4), 266–277. https://doi.org/10.1016/j.syapm.2009.04.001.
122. Wilkie, M. P., (1997). Mechanisms of ammonia excretion across fish gills. *Comparative Biochemistry and Physiology Part A: Physiology, 118*(1), 39–50. https://doi.org/10.1016/S0300-9629(96)00407-0.
123. Wuang, S. C., Khin, M. C., Chua, P. Q. D., & Luo, Y. D., (2016). Use of Spirulina biomass produced from treatment of aquaculture wastewater as agricultural fertilizers. *Algal Research, 15*, 59–64. https://doi.org/10.1016/j.algal.2016.02.009.
124. Yokoyama, H., & Ishihi, Y., (2010). Bioindicator and biofilter function of Ulva spp. (*Chlorophyta*) for dissolved inorganic nitrogen discharged from a coastal fish farm—Potential role in integrated multi-trophic aquaculture. *Aquaculture, 310*(1), 74–83. https://doi.org/10.1016/j.aquaculture.2010.10.018.
125. You, X., Shan, X., & Shi, Q., (2020). Research advances in the genomics and applications for molecular breeding of aquaculture animals. *Aquaculture, 526*, 735357. https://doi.org/10.1016/j.aquaculture.2020.735357.

126. Zang, C., Huang, S., Wu, M., Du, S., Scholz, M., Gao, F., & Dong, Y., (2011). Comparison of relationships between pH, dissolved oxygen and chlorophyll a for aquaculture and non-aquaculture waters. *Water, Air, & Soil Pollution, 219*(1), 157–174. https://doi.org/10.1007/s11270-010-0695-3.

127. Zappernick, N. E., (2020). *Techno-Economic Analysis of a Tilapia-Lettuce Aquaponics System*. The Ohio State University. Retrieved from https://etd.ohiolink.edu/apexprod/rws_olink/r/1501/10?clear=10&p10_accession_num=osu1610027069230629 (accessed on 25 June 2022).

128. Zupo, V., Graber, G., Kamel, S., Plichta, V., Granitzer, S., Gundacker, C., & Wittmann, K. J., (2019). Mercury accumulation in freshwater and marine fish from the wild and from aquaculture ponds. *Environmental Pollution, 255*, 112975. https://doi.org/10.1016/j.envpol.2019.112975.

# INDEX

**G**



**H**

**I**



**L**



**M**

**N**



**O**



**P**

T



U



V



W

**Z**